# REVOLUTIONS IN PHYSICS

# REVOLUTIONS IN PHYSICS

*Exploring the evolution and state of modern physics
and the possibilities that a new paradigm
holds for human civilization*

## Joseph P. Firmage

October 1, 2009

**To order additional copies of this book, contact:**
Xlibris Corporation
1-888-795-4274
www.Xlibris.com
Orders@Xlibris.com
99122

dedicated to the children of earth

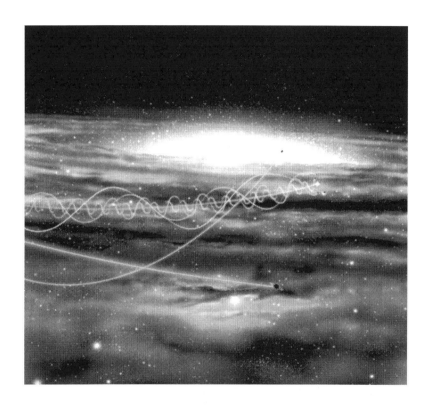

Artist's Rendering of Motions of Solar System
Bodies Relative to Center of Milky Way

# Chapter 1
# The Mystery of Mass in Motion

"A man that is of Copernicus' Opinion, that this Earth of ours is a
Planet, carry'd round and enlightn'd by the Sun, like the rest of them,
cannot but sometimes have a fancy . . . that the rest of the
Planets have their Dress and Furniture, nay and their
Inhabitants too as well as this Earth of ours . . .
But we were always apt to conclude, that 'twas in vain to
enquire after what Nature had been pleased to do there,
seeing there was no likelihood of ever coming to an end of the
Enquiry . . . but a while ago, thinking somewhat seriously on this
matter (not that I count my self quicker sighted than those great Men,
but that I had the happiness to live after most of them) me thoughts the
Enquiry was not so impracticable nor the way so stopt up
with Difficulties, but that there was very good room
left for probable Conjectures."[1]

—Christiaan Huygens

For many thousands of years we human beings have pondered the way the
Universe works. As the methodology of science formalized over millennia,
students of Nature organized maps of their ponderings into domains correlated
with the territories of the reality emoted in their senses. In biology we discover
the nature of living beings. Earth's composition and processes are revealed
by the discipline of geology. Chemistry is the lens through which we glimpse
the constituents and interactions of substances. We explore the depths of
the Cosmic ocean through the discipline of astronomy. In the past century,
the practice of science has blossomed around the globe, yielding dozens of

generalized and thousands of specialized domains of study, each one dedicated to the revelation of knowledge about a realm of Nature previously beyond the reach of human understanding.

Among the earliest disciplines of science to emerge was physics, through which we attempt to comprehend the commonalities shared by all of Nature's creations. As far into the mists of history as we are able to peer, we find human imaginations struggling to comprehend what the objects we sense are made of, and why they move and interact the way they do. Indeed, the grandest constructions of ancient human civilizations bear living testimony to the importance our ancestors placed on the mysteries of bodies in motion. The significance endowed by the ancients on the study of the clockwork of the Cosmos was well placed, for physics describes the irreducible elements and processes upon which all higher sciences are conceived. Today, human society is thoroughly dependent upon knowledge of physics for uncountable aspects of daily life. Knowledge of physics feeds us, clothes us, moves us about in land, sea, and airborne vehicles, powers our homes, cools us when hot and warms us when cold, and equips us with wondrous tools for computation, commerce, education, entertainment and communications.

In view of the reliance of science and society on the learnings of physics, the non-physicist might presume that the most fundamental questions in this most fundamental realm of science have been answered in clear terms. This is not so. While some physicists believe that what remains to be discovered is but a ribbon capable of tying together existing theories, other physicists are not at all so sure.

The two great theories of modern physics—quantum mechanics, which describes the small-scale workings of matter—and general relativity, which describes the large-scale motions of matter—are widely acknowledged by physicists to be in serious conflict with each other. As the prolific theoretical physicist Brian Greene points out, "The two theories underlying the tremendous progress of physics during the last hundred years . . . are mutually incompatible.

If you have not heard previously about this ferocious antagonism you may be wondering why. The answer is not hard to come by. In all but the most extreme situations, physicists study things that are either small and light (like atoms and their constituents) or things that are huge and heavy (like stars and galaxies), but not both. This means that they need use only quantum mechanics or general relativity and can, with a furtive glance, shrug off the barking admonitions of the other. For fifty years this approach has not quite been as blissful as ignorance, but it has been pretty close."[2]

Despite epic strides in physics achieved in the past century, fundamental questions remain unanswered—questions that stretch back to the origins of science and religion, questions that involve our deepest concepts of reality. The state of exploration of these questions deserves popular illumination, for the implications of possible answers are hard to overstate.

Imagine that a startlingly profound set of discoveries may spring out of answers to these questions. If these discoveries unfold in the direction suggested by research currently underway around the world, revealing that the "vacuum of space" is in fact an intense medium of energy, long-held assumptions regarding the limits of the technologies central to the infrastructure of civilization—the energy that powers all machines and the propulsion systems that enable us to move—will be swept away, and an astonishing array of ultra-efficient energy and transportation machines may begin to emerge.

If my speculations on the possibility of such technologies prove true, then our society may well be faced with the most intense period of evolution in the history of civilization. Such new categories of tools would be as profoundly transformative to 21st century civilization as fire was to ancient man: noninvasive fields capable of eliminating harmful bacteria and viruses by targeting their electromagnetic geometries; nonpolluting energy generators with no need for a grid; unlimited pure water from the ocean; silent, hovering vehicles that make congested freeways a stress of the past; and perhaps most compelling of all, spaceships that can carry us to the stars.

Fundamentally, the transformation we may face in coming years, rising from newly clear comprehension of the physics of our Nature, may include deeply profound insights into such questions as the existence and presence of other intelligent life in the Universe, the history of the Universe itself, and our future within it.

As the first step in exploring open questions in physics, let us survey the evolution of concepts essential to the epistemology and ontology of classical physics.

# Chapter 2
# Epistemological and Ontological
# Concepts in Classical Physics

"Throughout its long history in human thought, from its
early adumbrations in Neo-Platonic philosophy, its mystic and
still inarticulate presentation in theology, to its scientific
manifestation in the physics of Kepler and Newton, to its
carefully thought-out redefinitions in positivistic and
axiomatic formulations, up to its far-reaching modifications in
modern theories of physics—nowhere does science seem to get
full command and control over all the conceptual intricacies involved.
One has to admit that in spite of the concerted effort of physicists and
philosophers, mathematicians and logicians, no final
clarification of the concept of mass has been reached."[3]

—Max Jammer

At the crux of the conflict between general relativity and quantum mechanics
are the very concepts employed in physics to describe what physics describes:
concepts of space, time, force and mass. The mental pictures of these
metaphysically crucial constructs were seeded in the philosophies of ancient
Greece, took scientific root in the Renaissance, and have evolved increasingly
formal description in the subsequent centuries of the scientific revolution.

## Concepts of Space

Among the most basic of concepts in physics is that of 'space'. Over the
past three millennia, notions of space have progressively differentiated into

two general and competing ideas: (a) space is a dimensionality in which all things exist, and (b) space is a dimensionality of which all things are made. In the former case, space is usually viewed as an idealized, unbounded, three-dimensional void in which matter exists, moves and interacts. In the latter case, space is usually viewed as a kind of 'metric' or 'field', whose dimensions are not "flat" in the common sense we experience, but rather 'warped' to cause and reflect the ways matter moves and interacts. In this latter view, it is now often speculated that matter particles themselves are knots or strings in the metric or field of space (space-time actually, a subject we'll visit later).

Ancient Greek philosophers articulated among the earliest primitive expressions of ideas about space, across this spectrum of concepts. Tending towards notion (a), as noted physicist Max Jammer writes in his thorough historical review, Concepts of Space, "According to Aristotle, numbers were accredited with a kind of spatiality by the Pythagoreans: 'The Pythagoreans, too, asserted the existence of the void and declared that it enters into the heavens out of the limitless breath—regarding the heavens as breathing the very vacancy—which vacancy distinguishes natural objects.'"4 Describing other views of the period, Jammer continues, " . . . in the first atomistic conception of physical reality space was conceived as an empty extension without any influence on the motion of matter."5 Uncomfortable with this view, Aristotle advanced an idea closer to notion (b), that space was akin to the modern concept of a field of force: "Moreover the trends of the physical elements . . . show not only that locality or place is a reality but also that it exerts an active influence."6

Despite the insights revealed by early Greek philosophers on the subject, scientific concepts of space would not fundamentally advance for nearly two millennia.

Then and throughout the intervening centuries, continuing well into the modern scientific revolution, theology proved to be an important influence on the ideas of physics. The influence of theology on concepts of space is evidenced in associations of God and space (and light) throughout Eastern spiritual literature, Greek metaphysics, in the Divan of Lebid, the Kabbala and the Zohar, and later throughout the Renaissance, in the writings of d'Alembert, Maupertuis, More, Descartes, Kepler, Gassendi, Newton and many others. To a lesser extent this influence continues today, as can readily be seen in the titles of books from physicists on Barnes and Noble's shelf.

But in the 14th and 15th centuries, both Aristotelian thought and the inseparability of religion and science were confronted with new philosophies. Rising from the ranks of Jewish philosophy, Hasdai Crescas advanced one of them concerning the question of space. Jammer writes, "Crescas tried to show that motion is not dependent upon the existence of a medium. The first step in his argument is the assertion that weight and lightness are intrinsic qualities of the bodies and independent of any medium . . ."[7] The view of space as a true void came to be held by a substantial fraction of the natural philosophers pondering the question.

But it was Isaac Newton who punctuated this debate with powerful ideas that reverberate throughout physics to this day. Concerning space Newton wrote, "I do not define time, space, place, and motion, as being well known to all. . . . I must observe, that the common people conceive those quantities under no other notions but from the relation they bear to sensible objects. And thence arise certain prejudices, for the removing of which it will be convenient to distinguish them into absolute and relative, true and apparent, mathematical and common . . .

"Absolute space in its own nature, without relation to anything external, remains always similar and immovable. Relative space is some movable dimension or measure of the absolute space; which our senses determine by its position to bodies; and which is commonly taken for immovable space; such is the dimension of a subterraneous, an aerial, or celestial space, determined by its position in respect to the earth. Absolute and relative space are the same in figure and magnitude; but they do not remain always numerically the same. For if the earth, for instance, moves, a space of our air, which relatively and in respect of the earth remains always the same, will at one time be in one part of the absolute space into which the air passes; at another time it will be another part of the same, and so, absolutely understood, it will be continually changed."[8]

According to Jammer, "Newton introduces absolute and immutable space, of which relative space is only a measure. The final degree of accuracy, the ultimate truth, can be achieved only with reference to this absolute space." Newton believed that some notion of absolute space was an epistemological (and possibly ontological) necessity for his first law of motion—that every body in uniform motion remains in uniform motion unless acted upon by an outside force—to be meaningful. He believed that a reference system independent of purely relative measures was a prerequisite for his entire system of mechanics to hold.

As further and perhaps more compelling justification for this view, Newton described the famous water pail experiment. Consider a pail of water at rest. The surface of the water is flat. But when the pail is rotated about its vertical center, its friction causes the water to swirl, and the surface of the liquid assumes a paraboloidal form, rising near the edges and falling in the center of the pail. Newton correctly identified that rotational motion—revealing what we now call centrifugal force—identifies some special frame of reference, which he identified as absolute space. Newton went so far as to suggest that this special frame of reference would cause the same effect on a rotating system even if that system were the only object existent in the universe.

During these early days of the scientific revolution, scientists were becoming increasingly reluctant to repeat errors of the past in presuming 'special' frames of reference (e.g. an earth-centered universe) as the basis for any mechanism of Nature. Thus, Newton's claims of absolute space caused much debate and controversy in philosophy and physics. George Berkeley disagreed with Newton's assumption that space itself need be absolute in order to obtain the effects observed in rotating systems: "If we suppose the other bodies were annihilated and, for example, a globe were to exist alone, no motion could be conceived in it; so necessary is it that another body should be given by whose situation the motion should be understood to be determined . . . Then let two globes be conceived to exist and nothing corporeal besides them. Let forces then be conceived to be applied in some way; whatever we may understand by the application of forces, a circular motion of the two globes round a common center cannot be conceived by the imagination. Then let us suppose that the sky of the fixed stars is created; suddenly from the conception of the approach of the globes to different parts of that sky the motion will be conceived."9 In this statement, Berkeley presaged Mach's principle of inertia, which to this day, more or less, has sufficed as an acceptable explanation of the behavior of water in Newton's spinning bucket.

Leibniz joined the debate opposing Newton's idea of absolute space: "If motion is nothing but change of contact or immediate vicinity, it follows that it can never be determined which thing is moved. For as in astronomy the same phenomena are presented in different hypotheses, so it is always permissible to ascribe real motion to either one or other of those bodies which change among themselves vicinity or situation; so that one of these bodies being arbitrarily chosen as if at rest, or for a given reason moving in a given line, it may be geometrically determined what motion or rest must be ascribed to the others so that the given phenomena may appear. Hence if there is nothing in motion but this respective change, it follows that no reason is given in nature why motion must be ascribed to one rather than to others. The consequence of this will be that there is no real motion. Therefore in order that a thing can be said to be moved, we require not only its situation in respect to others, but also that the cause of change, the force or action, be in itself."10

Leibniz also held contempt for Newton's proclivities to correlate absolute space with God: "If the Space . . . void of all Bodies, is not altogether empty; what is it then full of? Is it full of extended Spirits perhaps, or immaterial Substances, capable of extending and contracting themselves; which move therein, and penetrate each other without any Inconveniency, as the Shadows of two Bodies penetrate one another upon the Surface of a Wall? Methinks I see the revival of the odd Imaginations of Dr. Henry More (otherwise a Learned and well-meaning Man), and of some Others, who fancied that those Spirits can make themselves impenetrable whenever they please. Nay, some have fancied, that Man in the State of Innocency, had also the Gift of Penetration; and that he became Solid,

Opaque, and Impenetrable by his Fall. Is it not overthrowing our Notions of Things, to make God have Parts, to make Spirits have Extension?"[11]

But as Jammer notes, "Nothing that Leibniz . . . had to say in criticism of Newton's concept of absolute space could prevent its acceptance . . . With the gradual acceptance of the Newtonian system, and as the rival Cartesian theories fell out of grace, Newton's concept of absolute space became a fundamental prerequisite of physical investigation."[12] Newton's ideas were followed by those of other natural philosophers, armed with new experiments seeming to suggest that space is some kind of physical medium. Indeed, while never believing in Newton's absolute space, Leibniz himself later became a proponent of space as a kind of medium.

In 1826, André-Marie Ampère published a groundbreaking study, summarizing the work of five years of research into the laws of the new science that he had named 'electrodynamics'. The theory formulated the forces that appear between the electrical currents in conductors. The experimental validation of Ampère's description of these forces was accomplished over the period 1832-1846, by Carl Friedrich Gauss's assistant and leading experimental collaborator, Wilhelm Weber. In 1831, Michael Faraday discovered the principle of induction, a concept independently developed by Joseph Henry. Faraday observed that a field of magnetic attraction and repulsion emerges around a conductor carrying electrical current, and conversely, that an electrical current can be induced in a conductor by moving a magnetic field relative to the conductor. Then in 1865, James Clerk Maxwell showed mathematically that magnets and electric currents should be able to produce traveling waves of electrical and magnetic energy, waves able to move through space on their own, free of the apparatus that produced them. Maxwell's theory was experimentally verified in 1888, and it was found that the waves produced traveled at the same speed as light. It was then no great leap to conclude that light itself was a form of 'electromagnetic' wave traveling through space.

Of great interest to physicists of the era was the observation by Faraday and Maxwell that the electromagnetic phenomena they observed appeared to be completely independent of any concept of absolute space. The relations between Faraday's electrical currents and magnetic fields were empirically observed to depend only upon relative motions, and the equations developed by Maxwell formalized this.

Nonetheless, the development of theories of electromagnetism suggested to many scientists that there must be some space medium through which these waves propagate. After all, if ocean waves travel through water, and if waves of sound travel through air, then how can light waves propagate through nothing? What does a light wave wave? As Ampère developed his theory of electrodynamics and Maxwell developed his theory of electromagnetism, Newtonian physicists postulated a kind of 'ether' as the medium that conveyed electromagnetic waves. Ether was held to be invisible and of such a nature that it did not interfere with the motions of bodies through space. The concept was intended to connect Newtonian mechanics with Ampère's and Maxwell's field theories.[13]

The development of concepts of the ether is a rich history unto itself, involving physicists of the highest stature, including William Thomson and Maxwell, but I shall not bother to expound upon it here, except in two specific respects: the importance of the issue to science and society, and the alleged refutation of the existence of any kind of ether. (Those interested in the details of the development of the ether concept should review the superb two-volume series A History of the Theories of Aether and Electricity by Sir Edmund Whittaker.)

Of the questions ever to confront science, the question of the nature of space ranks among the most significant. The entire apparatus of modern theoretical physics is hinged upon the answers to this and a few other equally basic, still elusive mysteries. As will become apparent in the discussions that follow, the nature of charges, atoms, electromagnetism, gravity, inertia, quantum phenomena, even the very definitions of 'time', 'force', 'energy' and 'motion' are intimately related to whatever it is that we call 'space'. So attempts to verify or refute the existence of the proposed 'ether' have been among the most important endeavors in the history of science.

Some of the early criticisms of the concept of an ether were based upon the absence of any apparent 'drag' on bodies in motion. If the ether were any kind of material substance, one would intuitively think that it would slow objects in motion as if they were moving through water or air. But it was easy to show that, if unobstructed by solid, liquid or gaseous matter, bodies in uniform motion do remain in such motion.

A more convincing argument against the existence of an ether was inferred from the results of the historic Michelson-Morley experiment. In 1887, the American physicists Albert Michelson and William Morley set out to determine the speed of the earth through the ether by measuring the relative speed of light traveling in different directions. They assumed that Earth could not be sitting still in the ether, so, considering Earth like a boat traveling through ethereal water, they assumed that waves would appear to travel at a different speed in the wake of the Earth as compared to the speed of waves traveling in the same direction the Earth was moving through the ether. To their shock, they observed no difference in the observed speed of light in any direction. It appeared that the relative speed of electromagnetic waves, unlike any other kind of wave previously observed, does not depend upon the motion of the observer—the observer always measures light to be arriving and departing at 299,793 kilometers per second. Their result puzzled physicists everywhere, for it was deeply counterintuitive to common sense.

Michelson and Morley's empirical result itself was later subjected to serious questions from Dayton Miller, among others, who claimed to have performed more sensitive experiments that did indeed detect motion of Earth relative to an ether, observing variance in the measured speed of light, with a particularly striking periodicity of measured velocity when plotted on sidereal time.[14] By the time these results were published, however, the concept of ether had been all but abandoned, and Miller's work was largely forgotten. More recent results remind us that this issue may not yet be closed.[15]

Setting that question aside, supported by the apparent relativity of electromagnetic phenomena, it was becoming increasingly apparent to most natural philosophers that the strictest interpretation of Newton's absolute space was not a requisite component of their theories. As Jammer notes, even before the emergence of electromagnetic theory, "Among the great French writers on mechanics, Lagrange, Laplace, and Poisson, none of them was much interested in the problem of absolute space. They all accepted the idea as a working hypothesis without worrying about its theoretical justification. In reading the introductions to their works, one discovers that they felt that science could very well dispense with general considerations about absolute space . . . the great success of Newtonian physics led to the paradoxical situation of the adherence to the concepts of absolute space, on the one hand, and their absence from practical physics, on the other."[16]

In 1885, Ludwig Lange proposed a way out of this paradox. He proposed the replacement of the notion of absolute space with the concept of the 'inertial system'. In essence, Lange replaced Newton's assertion that true motion occurs relative to absolute space with the less constraining assertion that motion occurs relative to a coordinate system in which Newton's law of inertia holds. He proposed that Newton's mechanics don't really depend upon the concept of absolute space, but rather simply depend upon a kind of space in which his laws remain valid. The idea represented a subtle but significant shift in the conceptual framework of physics.

This idea fit hand in glove with the contemporary work of other theorists, many of whom were in the midst of developing theories of geometry that leaped beyond Euclid, Cartan, and Descartes. Georg Riemann pioneered systems of mathematics and geometry in which n-dimensional 'spaces' could be conceived as being curved, or 'warped'. It was sensed that through such constructions, Newton's absolute maps of uniform space and time might be replaced by other maps of territories intrinsically at odds with our common sense of them. Through Riemannian geometries, Lange's suggestion that Newton's mechanics is valid, but with respect to a dimensionality other than absolute space and time, would ultimately find broad acceptance among 20th century physicists.

Ernst Mach soon presented a reasonable—though unverified—hypothesis to explain the tug of inertia on matter in all forms of accelerated motion, including Newton's bucket of water. He suggested that the gravitational attraction of all objects in the Universe on any one object is the reason why objects have inertia, why they resist acceleration. And since no experimental test had yet been devised to detect motion relative to absolute space or any kind of ether, the fixture underlying Newton's mechanics was increasingly open to reconsideration.

These circumstances helped set the stage for the emergence of Albert Einstein's theory of relativity—a hypothesis of epochal significance to physics' concept of space. But before we review Einstein's relativity, we shall examine the evolution of other basic concepts of physics that would soon interweave throughout his formulations.

## Concepts of Time

As elusive as concepts of space may be, 'time' seems a more mysterious quality of existence. One of the finest popular texts written on the subject is physicist Paul Davies' beautifully written volume, About Time. In its pages are found the wonderings of philosophers, poets, prophets and physicists on this mystery of eternity. Following are a few of my favorites:

> And likewise time cannot itself exist,
> But from the flight of things we get a sense of time . . .
> No man, we must confess, feels time itself,
> But only knows of time from flight or rest of things. [17]
> > —Lucretius

> > And the end and the beginning were always there
> > Before the beginning and after the end.
> > And all is always now. [8]
> > > —T.S. Eliot

> I see the Past, Present, and Future existing all at once, before me. [19]
> > —William Blake

The word Time came not from heaven but from the mouth of man.[20]

—John Wheeler

*The hands of the clock have stayed still at half past eleven for fifty years. It is always opening time in the Sailors Arms.* [21]

—Dylan Thomas

One of the interesting observations that can be made of the nature of time is that it has received less formal attention among scientists than the nature of space. Hans Reichenbach, whose philosophical expositions of physics are among the more respected in the field, writes, "Philosophy of science has examined the problems of time much less than the problems of space. Time has generally been considered as an ordering schema similar to, but simpler than, that of space, simpler because it has only one dimension . . . it has none of the difficulties resulting from multidimensionality."[22]

Another reason why the nature of time has received less formal examination than it may deserve is that we humans appear to have trouble forming a coherent mental picture of what it is that we should examine. This fact is plainly evident from a tour of concepts of time through the ages.

Plato opined that our sense of time is only partly real, an ephemeral experience of timeless forms which occupy eternity. Time is a "moving image of Eternity which remains forever at one . . . The past and the future are created species of time, which we unconsciously but wrongly transfer to the eternal essence."[23] Aristotle did not employ the notion of time per se in his physics. He understood its importance, but believed that time was, literally, motion. This is an interesting concept to which we will return later in this monograph, but the view of time as motion never gained much currency in the history of the subject, to this day. Plotinus, a third-century pagan, believed that to exist in time is to exist imperfectly. Pure Being must be characterized by the utter absence of any relation to time. He believed that time represents a prison for human beings, separating us from the divine realm—the true and absolute reality.[24]

This belief that the concept of 'being' exists outside of time became the established doctrine of many Christian thinkers, such as Augustine, Boethius and Anselm—a belief system that continues to this day. According to Augustine, time does not pass in the way we assume: "Your years are completely present to you all at one because they are at a permanent standstill. They do not move on, forced to give way before the advance of others, because they never pass at all . . . Your today is eternity."[25]

Many theologians and philosophers—and indeed countless ordinary people—report having experienced a kind of awakening of consciousness in relation to time. Physicist and Anglican bishop Ernest Barnes, in his 1929 Gifford Lectures, eloquently captures one such experience: "I remember that I was going to bathe from a stretch of shingle to which the few people who stayed in the village seldom went. Suddenly the noise of the insects was hushed. Time seemed to stop. A sense of infinite power and peace came upon me. I can best liken the combination of timelessness with amazing fullness of existence to the feeling one gets in watching the rim of a great silent fly-wheel or the unmoving surface of a deep, strongly-flowing river. Nothing happened: yet existence was completely full. All was clear."[26]

Eastern mystics believe that they have perfected practices that can induce these rapturous experiences routinely. The Tibetan Monk Lama Govinda describes his own experience thus: "The temporal sequence is converted into a simultaneous co-existence, the side-by-side existence of things into a state of mutual interpenetration . . . a living continuum in which time and space are integrated."[27]

Davies writes, "The Indian philosopher Ruth Reyna believes the Vedic sages 'had cosmic insights which modern man lacks . . . Theirs was the vision not of the present, but of the past, present, future, simultaneity, and No-Time.' Sankara, the eighth-century exponent of Advaita Vedanta, taught that Brahma—the Absolute—is perfect and eternal in the sense of absolute timelessness, and thus the temporal, though real within the world of human experience, has no ultimate reality . . . a truly timeless reality may be attained 'not in the sense of endless duration, but in the sense of completeness, requiring neither a before nor an after,' according to Reyna. 'It is in this astounding truth that time evaporates into unreality and Timelessness may be envisioned as the Real."[28]

Walter Ong, an expert on temporal symbolism, associates the reaching for a grander concept of time with an escape to the comfort of mythologies, and yet leaves the door open for a future rationalization of that dream with science: "Time poses many problems for man, not the least of which is that of irresistibility and irreversibility: man in time is moved willy-nilly and cannot recover a moment of the past. He is caught, carried on despite himself, and hence not a little terrified. Resort to mythologies, which associate temporal events with the atemporal, in effect disarms time, affording relief from its threat. This mythological flight from the ravages of time may at a later date be rationalized by various cyclic theories, which have haunted man's philosophizing from antiquity to the present."[29]

The theme that our day-to-day sense of time obscures its true nature is recurrent through both Western and Eastern thought, and many of us have the intuition that this is so. Every one of us experiences time in a very real but subjective manner. Time is intrinsic to our intuition of some kind of eternal existence. Yet we sense that things are somehow flowing from past to future, and we are in constant witness to births and deaths of bodies over time. So it seems hard for us to jibe the idea of an 'eternal now' with the experience of the ever-flowing sequence of change we sense through every waking moment.

Intellectually, we know that the measurement of time is a fundamental construct in almost all sciences, yet it is hard to identify a consensus mental picture of the nature of time, even among modern physicists.

As for classical physics, the crucial role of time was made fully apparent with the work of Newton. Similar to his views on space, he wrote, "absolute time, true and mathematical time, of itself, and from its own nature, flows equably without relation to anything external." Armed with the astounding success of his mechanics, Newton's views on the nature of time quickly ascended to become the belief structure of science as a whole. Central to Newton's notion was that material bodies move through space along predictable paths influenced by forces, all the while adhering to mathematical laws. This empirically-validated concept supported Newton's contention that time, like space, was absolute and, in itself, unchanging.

Many natural philosophers and theologians had some difficulty with the implications of this concept, for if all the motions of bodies were predictable, then where is the role for self-determination in the course of human life? While the question of determinism is outside the scope of this work, suffice to say that the debate remains unsettled even today, though the appearance of quantum mechanics has proposed interesting insights into this matter.

Upon his concept of an absolute 'flux' of time, Newton developed his theory of 'fluxions'—a branch of mathematics that became known as the calculus. With the growing precision of clocks and the success of Newton's theories, it became less fashionable to argue that time is merely an illusion, a mental construct created by humans in their failure to grasp eternity. Now time was seen at work within the very laws and mechanisms of the Universe.

Up to the end of the 19th century, while much was written on the subject of time, it is fair to say that the construct of time in physics remained reasonably consistent with Newton's philosophy. The scientific concept of time would not substantially change until the emergence of Einstein's theory of relativity, and then, as with concepts of space, all bets were off. Yet before Einstein and since, the closest science has come to defining the nature of time is by reference to the mechanism of the clock. Skipping ahead in our historical review, the view of modern physics is simply, and accurately, stated by physicist Brian Greene: "It is difficult to give an abstract definition of time—attempts to do so usually wind up invoking the word 'time' itself, or else go through linguistic contortions simply to avoid doing so. Rather than proceeding down that path, we can take a pragmatic viewpoint and define time to be that which is measured by clocks."[30]

## Concepts of Mass

Though 'mass' is the physics term that denotes the essential quality that makes matter matter, the question of the nature of mass does not often consciously bear on one's everyday life. The ever-relevant properties of mass—the way matter pushes back on you whenever you push it, through a property physicists call inertial mass, and the falling of matter to the ground, through a property physicists call gravitational mass—are so ordinary and ubiquitous that we seldom stop and think about them. The mystery of mass hasn't even held great relevance for the development of applied technologies throughout the $20^{th}$ century. Ever since Newton produced his great theories of mechanics, engineers have known most of the principles of mass they need to know in order to make things do what they do.

The nature of mass is nonetheless a very basic question, and as Jammer notes, has never been satisfactorily answered by science. It is therefore not surprising that, as he says, "Neither textbooks nor lecture courses seem to give a logically as well as scientifically unobjectionable presentation of the concept."[31] Physicist Herbert Jackson explained this circumstance, "Mass may be compared with an actor who appears on the stage in various disguises, but never as his true self . . . It may appear in the role of gravitational charge, or of inertia, or of energy, but nowhere does mass present itself to the senses as its unadorned self."[32]

The etymology of the word 'mass' is a useful starting point in the exploration of the development of the concept. The modern word was likely derived from the Latin 'massa', meaning a lump of dough or paste. Some have argued that the term has roots in the Hebrew word 'mazza' and the Greek 'maza' with meanings of bread or cake, or more broadly, food. Others have suggested that the Hebrew term was derived from 'macu', meaning to swell or extend in space. Jammer suggests that "the concept of mass originated from a logical analysis concerning the Eucharistic transubstantiation of the Holy Bread."[33] By the era of Middle English, it had evolved to mean a conglomeration or aggregation of bodies.

But scientific concepts of the weight of mass had other roots. The rise of commerce and the exchange of goods demanded means to compare the quantities of matter, by both weight and bulk. The use of balances to measure weight date back to prehistoric times, and perhaps as late as the 7th century B.C., the time-saving device of the coin was developed, enabling people to avoid repetitive weighing and otherwise intermediate the process of exchange. But these measures of matter were not comprehended for many centuries to come. Weight was not generally conceived as a dynamical process, but rather a quality like color or brittleness.[34]

It may have been this misconception that led Aristotle to his erroneous conclusions that the free-fall of a particle depends upon whether the particle is part of a big and bulky object or a small one. That and related debates concerning the manner in which bodies fall would not be settled until Galileo's famous experiment. But Aristotle's ideas on the reasons why matter moves the way it does were interesting nonetheless. Aristotle believed that motion is the resultant of two forces, one impelling and one resisting but both outside the body itself. Thus, Aristotle seems to have held the notion that the mass of matter is extrinsically sustained, but did not consider weight as a true measure of the quantity of matter. While this notion is not exactly true, Jammer makes what in my view is a questionable assertion that, "In modern terms, it may be said that Aristotle's dynamics is a logically consistent theory for motions either in a gravitational field or in a resisting medium; for motions in a vacuum (whose existence Aristotle does not acknowledge) and in the absence of gravitation his theory breaks down . . ."[35]

Lucretius' views were different than those of Aristotle: "Why do we find some things outweigh others of equal volume? If there is as much matter in a ball of wool as in one of lead, it is natural it should weigh as heavily, since it is the function of matter to press everything downwards, while it is the function of space on the other hand to remain weightless." Jammer observes that through this idea, "Weight is no longer an accidental property of matter but becomes a universal attribute of matter.[36]

In Neoplatonic thought, matter is conceived as a matrix of geometrical forms. Pointing towards future useful concepts of mass, Proclus asserts that the passivity or inertia of matter follows from its spatial extension: "For matter, as matter (soma), has no character save divisibility, which renders it capable of being acted upon, being in every part subject to division, and that to infinity in every part." Jammer observes, "Since extension, in Proclus's view, is tantamount to unlimited divisibility, something infinitely divisible is subject to external activity to an unlimited extent and consequently is only of a passive nature. Thus, according to Proclus, extension . . . gives rise to passivity or inertia."[37]

This notion found expression in the philosophy of Muslim society through pseudoepigraphic writings in an Arabic translation in the 9th century, and contributed to a neoplatonization of Aristotle in Muslim philosophy. This community of thought added an interesting assertion: "Rest is more in accord with the concept of matter than motion, because matter, though having six sides (i.e. three dimensions), cannot move simultaneously in all six directions; and a movement in one direction is not preferential to that in another; therefore immobility is characteristic of matter."[38]

One of the interesting mental pictures that seemed to follow from these ideas, and which gained much currency in philosophical and religious thought, was that matter was shapeless, clumsy and perhaps even ugly. Alanus ab Insulis, a 13th century author, in his encyclopedic poem Anticlaudian describes the clumsy nature of the "ugly matter", and speaks of its "sordid complexion" and "miserable deformation", but also of its "quest for a better look and a more graceful appearance." Jammer observes, "It is this characterization of matter that Johannes Kepler, the originator of our modern concept of inertial mass, had in mind when he wrote: 'All corporeal stuff or matter in the whole world has this virtue, or rather vice, that it is plump and clumsy to move itself from one place to another.'"[39]

The nature of matter received considerable attention in medieval thought, and helped form the basis for later scientific concepts of matter and mass. Avicenna identifies corporeal form with the predisposition of matter to assume spatial extension or three-dimensionality. Algazel asserted that corporeal form is the cohesiveness or massiveness of matter which constitutes the basis for the three-dimensionality of matter. Introducing an important subtlety into these ideas, Averroes suggested that corporeality is indeterminate three-dimensionality, but not the variable and measurable three-dimensionality of matter which he refers to as determinate three-dimensionality. He suggests that determinate dimensionality is capable of being increased or decreased, whereas indeterminate dimensionality is a characteristic of matter's existence—an ontological quality of matter.

According to Jammer, "This controversy concerning the nature of the corporeal form is of importance for our subject for several reasons. In the first place, it is an expression of a general tendency to find something that characterizes the essence of matter and yet is different from spatial extension. Secondly, the Averroistic concept of indeterminate dimensions, though slightly modified, becomes an important factor in Aegidius' definition of quantitas materiae, the first explicit definition of the concept of mass . . . This statement, although made by Averroes more or less in an incidental manner, is most remarkable. For it alludes to the possibility of conceiving the essence of matter in its dynamic behavior. Historically viewed, it is the earliest, although as yet highly inarticulate, expression of a dynamical concept of mass."[40]

With this concept in mind, we can see the brilliance of Aegidius Romanus, a disciple of Thomas Aquinas, in his formulation of quantitas materiae. As may become clearer in later sections of this monograph, the following discussion deserves careful thought, and extensive quotes from Jammer[41]:

"Aegidius's point of departure . . . [is the problem of] condensation and rarefaction . . . The difficulty [of determining matter by size versus density] disappears as soon as it is assumed that it is not one and the same quantity that is under discussion. In order words, if we assume that there are two different kinds of quantities, then one kind can sustain . . . the variation of the other kind . . . and no logical inconsistency is involved . . . This theory of duplex quantitas thus explains condensation—even in the absence of substance—as a ratio between the two quantities that are called determinate and indeterminate dimensions, the former corresponding to volume and the latter to what was later called quantitas materiae."

Aegidius wrote, "It should be understood that in the matter of the bread and the wine as well as in all earthly matter there are two quantities and two kinds of dimensions: determinate and indeterminate dimensions. For matter is so and so much and occupies such and such a volume . . . If it can be shown that it is not the same quantity by which matter is so and so much and by which it has such and such a volume, and, on the other hand, if we can state the quantity in virtue of which matter is so and so much precedes the quantity in virtue of which it occupies such and such a volume and that in the first kind of quantity, as in a subject, the second kind of quantity is anchored, then it is easy."

Jammer continues, "Aegidius thus accepts the Averroistic terminology of determinate and indeterminate dimensions, but their meaning is slightly different: dimensiones indeterminatae is now the name for quantity of matter (quantitas materiae) and dimensiones determinatae the name for volume, the measurable, determinable spatial quantification. This distinction, argues Aegidius, follows from the fact that the variation of one of them does not imply a variation of the other, as, for instance, in the process of rarefaction in which the determinate dimensions increase while the quantity of matter remains invariant . . . Aegidius thus postulated the existence of a quantitative measure of matter that is different than volume determination . . . The Aegidian innovation is, indeed, a radically new conception, a new 'dimension' in the modern technical sense of the word . . . And yet, in spite of Aegidian's clear presentation of his ideas, his innovation gained little recognition among the Schoolmen. In fact, Aegidius himself, after 1289, renounced his ideas . . . Thus Thomas of Sutton, a member of the English Thomistic school, at the end of the thirteenth century criticized Aegidius's theory . . . and proclaimed the essential identity of the two quantities."

Eulogizing a concept of mass that I believe holds great import for the debates of modern physics, Jammer concludes this episode, "Thus a concept which, owing to the inherent lack of operational significance would in any case have proved itself unacceptable for modern science."

Nonetheless, in following centuries the concept of mass would leap forward, both in terms of theory and empirical utility. Buridan described the proportionality between impetus and quantity of matter. Galileo demonstrated that objects free-fall at the same velocity. Descartes conceptualized momentum and its conservation. But two pioneers of physics' systematization of the concept of mass stand out: Johannes Kepler and Isaac Newton.

Kepler began the process of equating force with resistance to a change in ongoing motion, marking the formal introduction of inertial mass into the apparatus of physics. In his pioneering studies of the motions of planets, he states: "If the matter of celestial bodies were not endowed with inertia, something similar to weight, no force would be needed for their movement from place to place; the smallest motive force would suffice to impart to them an infinite velocity. Since, however, the periods of planetary revolutions take up definite times, some longer and others shorter, it is clear that matter must have inertia which accounts for these differences." "Planetary bodies . . . are not to be considered as mathematical points but obviously as material bodies endowed with something like weight or rather an intrinsic faculty of resistance to motion which is determined by the volume of the body and the density of its matter."[42]

Hand in hand with achieving this understanding of matter in motion, and after years of painful and often depressing failures, Kepler successfully answered one of the great mysteries of Nature: the patterns in which planets orbit the sun. His mathematics of planetary motions eventually proved correct, and Ptolemy's aged and brittle theory of 'epicycles' was swept away.

Soon after, another of the greatest minds ever to live presented profoundly compelling new theories, tying the threads of force, inertia, and motion together with beautiful elegance and simplicity. In Isaac Newton's Principia, we read:

"This force is always proportional to the body (suo corpori) whose force it is and differs nothing from the inactivity of the mass (inertia massae), but in our manner of conceiving it. A body, from the inert nature of matter, is not without difficulty put out of its state of rest or motion. Upon which account, this vis insita may, by a most significant name, be called inertia (vis inertiae) or force of inactivity. But a body only exerts this force when another force, impressed upon it, endeavors to change its condition; and the exercise of this force may be considered as both resistance (resistentia) and impulse (impetus); it is resistance so far as the body, for maintaining its present state, opposes the force impressed; it is impulse so far as the body, by not easily giving way to the impressed force of another, endeavors to change the state of that other. Resistance is usually ascribed to bodies at rest, and impulse to those in motion; but motion and rest, as commonly conceived, are only relatively distinguished; nor are those bodies always truly at rest, which commonly are taken to be so."[43]

It is from this succinct line of reasoning that Newton's three laws of motion naturally follow. Jammer makes the questionable observation that Newton's statement makes no explicit use of the concept of mass. But Jammer also offers a more reasonable, and important, clarification of this assertion, that "for Newton the notion of density was primary and anterior to the concept of mass." The interesting implications of this possibility were at least glimpsed by Rosenberger, Bloch, and Crew. The latter noted that "On such a system, is it both natural and logically permissible to define mass in terms of density." It might have been better said that one could define the quality of mass as a density of something more fundamental. Enrique went further to suggest that Newton was seeking both a qualitative and quantitative concept of mass in terms of density times volume. [44]

Interpretations aside, Newton's work was utterly historic. In Principia he set forth interrelations for most of the basic concepts of physics, in a simple and rugged construct. Using these principles, Newton pioneered the first effective law of gravitation, describing with nearly perfect accuracy the manner in which objects attract one another. As empirical validations continued to appear, Newton's mechanics became recognized for what it was: a framework for the systematic development of physics, and by extension, of all physical science. However, as effective as his reasoning was in describing the mechanics of bodies in motion, Newton's laws represented only relations among 'force', 'inertia', and 'change in motion'. They did not define the nature of force or mass or acceleration, but rather clarified their relationship, and in the meantime, gave theoreticians and engineers new tools to explore and work with Nature.

In part building upon Newton's work, and in part debating alternative physical interpretations of his equations, other bright lights carried forward the quest for concise definitions of mass.

Of particular note was Leibniz's theory of matter. Leibniz's concept of mass was considered by some to be quite different than that of Newton, however it is possible that future discoveries may prove this contention less true than originally believed. Leibniz wrote, "Primary matter is mass itself in which there is nothing but extension and antitypy [repulsion?] or impenetrability." Jammer elaborates, "Leibniz's theory of matter can be understood only if viewed against his doctrine of monads [a single kind of irreducible constituent whose collections comprise all things]. Primary matter (material prima), according to Leibniz, is not body but intrinsic to the 'being a monad.' Secondary matter (material secunda), on the other hand, does not refer to the monads themselves but belongs to a group, or cluster, of monads; it is founded on the ontological relation between the subordinate monads, reflecting by their nature with lower degrees of clearness, are also the more passive ones. These, in turn, have also secondary matter through the subordination of other monads to them, and so on ad infinitum . . . Leibniz now applies the same terms of primary and secondary matter also to the world of perception, the world as the object of physical research and not of metaphysical scrutiny. Primary matter, or mass . . . , is here an abstraction; it is body conceived as merely occupying space and preventing other bodies from occupying the same space. Extension and antitypy, a favorite term with Leibniz for impenetrability, are thus the attributes of material prima." Separately, Leibniz wrote, " . . . primary matter . . . is not a substance, nor even an aggregate of substances, but something incomplete. Secondary matter, or mass, is not one substance, but a plurality of substances." [45]

Though they sketched many vital notions, foreshadowing the structure of matter as comprised of fundamental entities assembled into constellations of larger entities (charges within atoms within molecules), Leibniz's concepts of matter gained little currency, perhaps to the directional misfortune of physics. But his explorations of the subject yielded at least two lasting impressions. First, Leibniz believed that a physical body could not be understood as extension alone. He believed that some other quality must belong to bodies by which their extension is physically meaningful, by which "a big body is more difficult to be set in motion than a small body." Leibniz believed that geometry in itself could not account for the spatio-temporal behavior of interacting bodies.

Second, Leibniz's implication that mass is by nature a kind of force gave urgency to others who believed that this could not be so. Among them was Immanuel Kant. For both Newton and Leibniz, the concept of inertial force carried by mass was an ontological necessity. While Kant shared this view for some time, he gradually came to reject it. Jammer writes, "in the metaphysical foundations of natural science . . . Kant rejects Newton's concept of vis inertiae altogether . . . As only 'motion can oppose motion, but not rest,' it is not the inertia of matter, its incapability to move itself, that produces resistance to the moving force. A force which by itself does not cause motion, but only resistance, is 'a word without any meaning.' The concept of a vis inertiae has to be abandoned in natural science, Kant concludes, not merely because of its paradoxical appellation, but because of the misconceptions which the term implies . . . Instead of the 'force of inertia' Kant postulates the 'law of inertia' as corresponding to the category of causality: every change in the state of motion has an external cause . . . Kant's elimination of the metaphysical vis insita or vis inertiae prepared the way for [other approaches] to the concept of mass."46 Kant's writings held substantial influence among natural philosophers of the era, and whether his appraisal of Newton's and Leibniz's concepts of inertia was correct remains, in my view, very much an open question.

The first rugged mathematical concept of mass was articulated by Leonard Euler in his Mechanica, a text devoted to the task of constructing a rigorous logical foundation for Newton's mechanics. The effort proved most fruitful, for it made clear the mathematical interpretation of Newton's concept of mass: the ratio of force to acceleration, expressed in the famous equation, f=ma. This equation is widely considered to be physics' first concise definition of mass. For the purposes of empirical measurement within the realm of classical mechanics, mass defined as the ratio of force to acceleration suffices. Whether that definition completely fulfills the ontological needs of modern theoretical physics is another question.

Leveraging the later emergence of non-Euclidean geometry, and in tandem with the freedom thus glimpsed to question the ontological nature of force, another concept of mass was articulated by Mach in 1868. Mach suggested that a purely 'kinematic' description of mass was possible, free of any dependence upon what in Newtonian physics would be called a field of force. [47] Mach's arguments (too complex for useful expression in this context) were met with mixed reactions. His constructions were apparently mathematically valid, though they depended upon the ability of bodies to attract and repel each other by some mechanism. Thus, Mach's approach depended upon either a construct of force, or upon a non-Euclidean metric—a construct that Einstein would later adopt as the foundation of his history-making theory of gravity.

Another very promising concept of mass emerged through the work of Ampere, Maxwell, Thompson, and others. It was in 1881 that Thompson envisaged the possibility that inertia might be reduced to an electromagnetic effect—an effect of the light-like forces that pervade all space in the Universe. He examined the electrodynamics of charged bodies moving in a medium of a specific inductive capacity (a specific 'dielectric constant'). While defects were identified in his theoretical approach, its philosophy showed promise. Heaviside tendered an improved version of the theory which appeared to suggest a mechanism akin to inertia. Jammer writes, "The publication of Heaviside's article marked the beginning of an animated competition between the science of mechanics and the science of electromagnetism for the primacy in physics. The era of mechanical interpretations of electromagnetic phenomena initiated by William Thomson (Lord Kelvin) and Maxwell in their search for mechanical models of the ether was still at its peak. The scientific journals of the [eighteen-]nineties were flooded with articles that tried to reduce electromagnetism to mechanics or hydrodynamics. The belief that all forces in nature are ultimately only different manifestations of the same fundamental power inspired many scientists to search for such principles of unification."[48]

The growing body of research supporting an electromagnetic basis for mass included Korn's mechanical theory of the electromagnetic field, Wien's efforts to explain the laws of mechanics through electromagnetic equations, Abraham's explorations of the electromagnetic nature of inertial mass, and Kaufmann's experimental tests of this hypothesis on electrons. The latter's experiments on the deflection of electrons by simultaneous electric and magnetic fields—establishing an apparent relation between velocity and electron mass—were sufficiently convincing for Kaufmann, Abraham, and others, including Hendrik Lorentz, to conclude that the mass of an electron originates in the electromagnetic field. Boltzmann nicely characterized the parsimony of the approach thusly: " . . . the advantage of deriving the whole science of mechanics from conceptions which anyhow are indispensable for the explanation of electromagnetism would be as important as if conversely electromagnetic phenomena were explained on the basis of mechanics. May the former succeed."[49]

According to Jammer, "The program of the electromagnetic conception of mass was now fully established: once ponderable atoms and molecules had been reduced to positive and negative charges and their inertial behavior had been explained on the basis of electrodynamics, an extension and generalization of this approach had to be found for molecular and gravitational forces. The whole universe of physics would then amount to merely positive and negative charges and their magnetic fields, all processes in nature would be reduced to convection currents and their radiations, and the 'stuff" of the world would have been stripped of its material substantiality . . .

"For the development of the concept of mass and thus for the development of physical theory in general, the electromagnetic theory of matter was of decisive importance. Till its advent, physicists and philosophers, on the whole, adhered to what was called the substantial concept of physical reality. A physical body, according to this view, is first of all what it is: only on the basis of its intrinsic, invariable, and permanent nature, of which mass was the physical expression and inertial mass the quantitative measure, did it act as it did. The electromagnetic concept, now, proposed to deprive matter of this intrinsic nature, of its substantial mass. Although charge, to some extent at least, fulfills the function of mass, the real field of physical activity is not the bodies but,

as Maxwell and Poynting have shown, the surrounding medium. The field is the seat of the energy, and matter ceases to be the capricious dictator of physical events."[50]

All in all, it was a most promising trajectory. Unfortunately, the early enthusiasm for this concept of mass began to fade as theoreticians appeared unable to carry out successful generalizations for constituents of matter other than electrons. But perhaps the greater reason explaining why the electromagnetic concept of mass fell from favor was that the key empirical evidence supporting it—the experimental discovery of velocity-dependence of electron mass—would soon find a different interpretation in Einstein's theory of relativity.

## Concepts of Force

The underlying physical nature of 'force' is among the most important open questions in the foundations of physics. As will become evident later in this monograph, the answer to this question will significantly clarify the nature of the Cosmos from which we spring and our future within it. To begin to understand why, let us take a more lengthy tour through the rich history of ideas on this vital question.

The pre-scientific notion of force remains the common sense concept most of us think of today: force is a 'push' or a 'pull' subjectively sensed in the interactions of bodies. In everyday use, concepts such as 'strength', 'power', and 'effort' are often used interchangeably with the concept of force. But the concept developed very early into something far more than a strictly subjective sensation. As Jammer writes in his exhaustive Concepts of Force, "The injection of our personal experience into the external environment, characteristic of the animistic stage in the intellectual growth of mankind, led to a vast generalization of the concept of force: trees, rivers, clouds, and stones were endowed with force and were regarded as centers of power. For what is active was thought to be alive, and an object, animal or material, being alive, was conceived as having within it the same sort of force that man recognized in himself. Moreover, things of nature that seemed to be powerful were not only endowed with an antropopathic nature; they also became objects of fear and reverence . . . The familiar expression of 'forces of nature' is still reminiscent of this outlook on nature."[51]

Over time, this panphysical conception came to describe the divine spirit or god of greatest power. Jammer continues, "The abstract concept of force, as a notion of divinity, can be traced back in ancient Egypt to the nineteenth dynasty and plays an important role in later texts of demotic literature . . . Wilhelm Spiegelberg in his paper on 'The Egyptian divinity of 'force" says explicitly: 'This abstract concept of 'force' with its attribute 'divine' as stated explicitly in the inscription was conceived as a personal deity.'" The relation of notions of force and God is also found throughout the Bible, as manifested in frequent associations of his name with 'might', 'power', 'strength' and 'vigor'. Some etymologists suggest "that one of the holy names, 'shaddai', derives from the Semitic root 'shadda', meaning to have great power, great force, and translated in the Septuagint as 'pantokrator' ('omnipotens')." The concept of force appears closely related to religious ideas throughout the records of ancient civilizations, maintaining this connection in the Platonic doctrine of force as an emanation of the world soul.[52]

Slowly, concepts of force began to take on clearer pre-scientific meanings. In Greek philosophy, early cosmologists believed Nature to be a living being, self-moving and giving birth to individual things. Heraclitus conceived of the intrinsic dynamism in Nature as opposing tensions, antagonistic forces whose appearance of stability is only relative, or even illusory. Empedocles saw force in the nature of 'love' and 'strife', the regulation of which yields motion. He believed force to be extended in space and corporeal. Anaxagoras perceived force as external to matter—a kind of world-forming spirit of the universe—whose dynamic ordering he termed 'mind'. For him, force is a kind of fluid substance different from all other material things. Plato interpreted the operation of Empedocles' forces of 'love' and 'strife' as sequences of attraction and repulsion, whereas he believed Heraclitus' concept held that both forces operate simultaneously. Plato's concept of force was rooted in his metaphysical doctrine of 'being' as producing change, and in turn affected by the changes caused by other beings. He wrote, "My notion would be, that anything which possesses any sort of power to affect another, or to be affected by another, if only for a single moment . . . has real existence." Jammer writes, "Plato's conception of force as something intrinsic in matter because matter has soul conforms with his view that the mutable world is composed of one single universal 'this' susceptible of many differentiated 'suches'; in principle, it is one substance or one substratum and one universal receptacle which manifests itself through forces and forms as different aspects in the actual 'this.'"[53]

Jammer continues, "The term generally used by Plato to denote this idea of force is the word dynamis. It is the noun corresponding to the verb dynastai, which means 'to be able,' 'to be capable.' . . . In contrast to our modern terms 'force,' 'power,' 'activity,' the Greek word dynamis signifies therefore not only transitive action or transeunt activity, but also passive susceptibility and receptibility . . . Heat and cold, chemical activity, botanical functions, hardness, and light are spoken of as dynameis." It was from this context that Aristotle chose dynamis as a technical term for any kind of push or pull, upon which he proceeded to define his laws of motion of bodies through a resistive medium.[54]

Concerning the concept of force, Theophrastus expresses the spiritually rich intuitions of philosophers of the era in a particularly beautiful poem:

For even as Love and Hate were strong of yore
They shall have their hereafter; nor I think
Shall endless Age be emptied of these twain.

Now grows
The One from many into being, now
Even from the One disparting come the Many,
Fire, Water, Earth and awful heights of Air;
And shut from them apart, the deadly Strife
In equipoise, and Love within their midst
In all her being in length and breadth the same.
Behold her now with mind, and sit not there
With eyes astonished, for 'tis she inborn
Abides established in the limbs of men.
Through her they cherish thoughts of love, through her
Perfect the works of concord, calling her
By name Delight or Aphrodite clear.
She speeds revolving in the elements,
But this no mortal man hath ever learned—
Hear thou the undelusive course of proof:
Behold those elements own equal strength
And equal origin; each rules its task;

And unto each its primal mode; and each
Prevailing conquers with revolving time.
And more than these there is no birth nor end;
For were they wasted ever and evermore,
They were no longer, and the great
All were then How to be plenished and from what far coast?
And how, besides, might they to ruin come,
Since nothing lives that empty is of them?
No, these are all, and, as they course along
Through one another, now this, now that is born—
And so forever down Eternity.[55]

These concepts became the doctrines shared by almost all ecclesiastical authorities during the Middle Ages. Thomas Aquinas considered Aristotle's conceptions "clearer and more sound" than those of any other approach. Jammer characterizes Aquinas' view: " . . . it is contended that no motion occurs without a mover; to avoid an infinite regress of movers, a first mover must be assumed who is himself unmoved. Since the motion under discussion is an eternal motion, the force, as the cause of this motion, must be an infinite force, and consequently cannot be originated by a corporeal being that possesses only finite qualities."[56] Muslim science was similarly founded upon Aristotelian conceptions of force.

Roger Bacon expanded upon these lines of reasoning in asserting that the transmission of force is a kind of chain reaction that successively energizes both the medium and the body. His notion of force as "species" held it as something corporeal, its corporeal nature identical with the nature of the medium. As Jammer writes, "The attraction of iron toward the magnet is frequently explained in the literature of the Middle Ages by

Bacon's conception of species: the magnet evokes in its environment a species magnetica that spreads spherically through the medium, multiplies itself from one portion of the iron to the adjacent one until the iron receives in all of its parts the magnetic quality of tending to become united with the magnet, and . . . gets subject to local motion. The action at a distance is reduced thereby to a chain of contiguous contact-processes. By a similar modus operandi the tendency of a gravis to its natural place is explained."[57]

But during this era, other ideas were brewing. William of Occam threw out the concepts of species and explicitly acknowledged the concept of action at a distance. He replaced Aristotelian principles of immediate contact between mover and moved with the notion that simultas virtualis is sufficient to maintain continuity of motion. Though his notion did not immediately gain much acceptance, it presaged a major change in physics' concept of the actions ascribed to force.

Modern science is widely characterized to have its inception in subsequent developments of the concept of force. Pierre Duhem declares this moment to have occurred thusly: "If one wishes to draw the line of separation between the realm of ancient and modern science, it must be drawn at the instant when Jean Buridan conceived his theory of momentum, when he gave up the idea that stars are kept in motion by certain divine intelligences, and when he proclaimed that both celestial and earthly motions are subject to the same mechanical laws." Buridan wrote, "Nowhere does one read in the Bible that there exist intelligences charged with communicating to the celestial spheres their proper movements; it is therefore permitted to show that there is no need to suppose the existence of such intelligences. One can say that God, when creating the world, has moved, as he pleased, each of the celestial orbits; he has given to each of them an impetus which kept them moving since then . . ."58 Indeed, it was the examination of planetary orbits that marked a decisive advance in the scientific concept of force, through Johannes Kepler's attempt to explain the motions of worlds by quantifying force.

In examining the work of observers who had correlated the motion of the moon with the tides of oceans, Kepler glimpsed a general principle at work: as water is influenced by the moon, as stones approach the Earth, so must the Earth approach the sun. He thus speculated that gravity is a "passivity" rather than an "activity", and is inherent in all bodies: "If two stones were removed to some place in the universe, in propinquity to each other, but outside the sphere of force of a third cognate body, the two stones, like magnetic bodies, would come together at some intermediate place, each approaching the other through a distance proportional to the mass (moles) of the other." Kepler expressed his conviction that this process is subject to mathematical rules when he said, "For we see that these motions take place in space and time and this virtue emanates and diffuses through the spaces of the universe, which are all mathematical conceptions. From this it follows that this virtue is subject also to other mathematical 'necessities.'"59

Kepler went on to observe an inverse relation between orbital speed and distance, asserting thus that gravity is proportional to the reciprocal of the distance. Although his math on this point was in error, he thus transformed pre-scientific concepts of force into an operational relation of physics. At the same time, he found no conflict between his mathematizations and his ontology, "If you substitute for the word 'soul' the word 'force,' you have the very principle on which the celestial physics of the treatise on Mars . . . is based . . . Formerly I believed that the cause of the planetary motion is a soul . . . But when I realized that these motive causes attenuate with the distance from the sun, I came to the conclusion that this force is something corporeal, if not so properly, at least in a certain sense."[60]

Galileo focused his attention on questions of motion more than questions of force, and cautions against premature conclusions as to the nature of the latter: "The present does not seem to be the proper time to investigate the cause of acceleration of natural motion concerning which various opinions have been expressed by various philosophers, some explaining it by attraction to the center, others to repulsion between the very small parts of the body, while still others attribute it to a certain stress in the surrounding medium which closes in behind the falling body and drives it from one of its positions to another. Now, all these fantasies, and others too, ought be examined; but it is not really worth while. At present it is the purpose of our Author merely to investigate and to demonstrate some of the properties of accelerated motion (whatever the cause of this acceleration may be)."[61]

This approach—examining the empirical patterns of apparent motions of materially-perceivable bodies—is referred to in modern physics as the 'positivistic' approach. As we shall see in later sections, Galileo's quasi-positivistic replacement of the question 'how' with 'how much' foreshadowed the predominant method of theoretical development employed in 20th century physics.

Galileo came close to the classical determination of force, and to a unified concept of static and dynamic forces, in writing: "It is clear that the impelling force acting on a body in a descent is equal to the resistance or least force . . . sufficient to hold it at rest. On order to measure this force and resistance . . . I propose to use the weight of another body." He lacked only a clearer concept of mass. As Jammer writes, "Galileo made it clear that its essence was hidden from him . . . left as an unanswered question. 'You should say that everyone knows that it is called gravity, and I do not question you about the name but about the essence of the thing. Of this you know not a tittle more than you know the essence of the mover of the stars in gyration, unless it be the name that has been put to the former and made familiar and domestic by the many experiences that we have of it every hour in the day.'

"With the recognition of the law of inertia, formulated already as early as 1585 by Benedetti as part of his impetus theory and presupposed by Galileo in his analysis of accelerated motion, or with the acceptance of the principle of conservation of motion, as announced by Isaac Beeckmann and Rene Descartes, two alternative possibilities presented themselves: either to conceive force as the cause of change of motion, or to abolish the notion of force altogether."[62]

Advancing the latter alternative Descartes rejected the notion of force altogether. On the grounds of the principle of inertia, he thought it might be possible to eliminate force as a separate physical concept. Jammer continues, "All physical phenomena, he contended, are to be deduced from only two fundamental kinematic [force-free] assumptions: the law of conservation of quantity of motion—which for him was not a corollary of the principle of inertia but its real physical content—and his theory of swirling ethereal vortices. For, rejecting any possibility of an action at a distance, Descartes constructs his vortex theory to account for the remote heavenly motions . . . The concept of force, in Descartes's view, had no place in his physics, which was to employ exclusively mathematical conceptions. [He wrote,] 'I do not accept . . . or desire any other principle in Physics than in Geometry or abstract mathematics, because all phenomena of nature may be explained by their means, and a sure demonstration can be given of them.' A geometrization of physics—this was Descartes's program before classical mechanics was born. It was a program too difficult, even for such an intellectual giant as Descartes . . . [But] Descartes's doctrine was regarded as an ad hoc theory which, contrary to its proclamation, had to smuggle in the concept of force under the cover of pressure or in another form."[63]

Another conception of motions in gravity was tendered by Giovanni Alphonso Borelli, which construed the planetary orbits as positions of equilibrium of opposing forces. This compelling idea was furthered by the work of Christiaan Huygens, who solved the mathematics of it in his Horologium oscillatorium. He mathematized the centrifugal force—the nature of which he did not comprehend—in action against the gravitational force, yielding the proper velocity-distance relation of orbiting bodies.

But, as reflected in our examinations of other fundamental concepts of physics, it would be Isaac Newton who would lay the foundation for the golden age of classical physics. Newton's laws of motion gave birth to a thoroughly useful mathematization of force that would be appraised as perfectly accurate through centuries of experimentation to come.

Jammer writes, "Much has been written on Newton's concept of gravitation but next to nothing on his concept of force in general . . . Newton's general considerations about force are methodologically related to his study of gravitation because the problem of a dynamical explanation of planetary motions to account for Kepler's three laws was the question of the hour." Stressing the importance of the question, Newton wrote: "I offer [Principia] as the mathematical principles of philosophy, for the whole burden of philosophy seems to consist in this—from the phenomena of motions to investigate the forces of nature, and then from these forces to demonstrate the other phenomena."[64]

Newton believed that force was an inherent quality of matter, latent so long as no other force is impressed upon the body: "This vis insita, or innate force of matter, is a power of resisting by which every body, as much as in it lies, continues in its present state, whether it be of rest, or of moving uniformly forwards in a right line . . . [Force] is resistance so far as the body, for maintaining its present state, opposes the force impressed; it is impulse so far as the body, by not easily giving way to the impressed force of another, endeavors to change the state of that other." This difference may be only an apparent one, since motion and rest are only relatively distinguished, as Newton himself seemed to suggest. In his examination of Newton's statements in this regard, Jammer makes a questionable assertion which I will discuss later: "Clearly, in this definition, force is not conceived as a cause of motion or acceleration."[65]

Even if the difference between impressed force and innate force were only in appearances, Newton specified three distinctions that were nonetheless useful to analysis (even if they prove not entirely accurate). First, impressed force is pure action, 'transeunt' in character; second, it no longer remains in the body once the action is over; and third, whereas inertia was innate to matter and not further reducible, impressed force could have many different origins, such as percussion, pressure, and centripetal force. His characterization of inertial force as innate and irreducible notwithstanding, Newton, like his predecessors, apparently considered the ultimate nature of force an open question, "For I here design only to give a mathematical notion of those forces without considering their physical causes and seats."[66]

However, qualitative concepts are far from absent in Newton's writings. In 1675 he offers some interesting concepts on the question of force: "It is to be supposed therein, that there is an aetherial medium, much of the same constitution with air, but far rarer, subtiler, and more strongly elastic . . . For the electric and magnetic effluvia, and the gravitating principle, seem to argue such variety. Perhaps the whole frame of nature may be nothing but various contextures of some certain aetherial spirits or vapours . . ." On the question of the nature of gravitational force specifically, Newton ventures further. In 1678 he wrote, "I shall set down one conjecture more . . . it is about the cause of gravity. For this end I will suppose aether to consist of parts differing from one another in subtility by indefinite degrees: that in the pores of bodies, there is less of the grosser aether in proportion to the finer, than in open spaces; and consequently, that in the great body of the earth there is much less of the grosser aether, in proportion to the finer, than in the regions of the air; and that . . . from the top of the air to the surface of the earth, and again from the surface of the earth to the center thereof, the aether is insensibly finer and finer. Imagine, now, any body suspended in the air, or lying on the earth; and the aether being, by the hypothesis, grosser in the pores which are in the upper parts of the aether, being less apt to be lodged in those pores, than the finer aether below; it will endeavor to get out, and give way to the finer aether below, which cannot be, without the bodies descending to make room above for it to go out into."[67]

More generally, and in response to Bentley who asserted that Newton's gravitational force was of a divine mind and above mechanism, Newton emphatically asserted: "It is inconceivable, that inanimate brute matter, should, without the mediation of something else, which is not material, operate upon and affect other matter without mutual contact, as it must be, if gravitation, in the sense of Epicurus, be essential and inherent in it. And this is one reason why I desired you would not ascribe innate gravity to me. That gravity should be innate, inherent, and essential to matter, so that one body may act upon another at a distance through a vacuum, without the mediation of any thing else, by and through which their action and force may be conveyed from one to another, is to me so great an absurdity, that I believe no man, who has in philosophical matters a competent faculty of thinking, can ever fall into it. Gravity must be caused by an agent acting constantly according to certain laws; but whether this agent be material or immaterial, I have left to the consideration of my readers."[68]

In summing up his qualitative concept of gravity, Newton forecloses only one possibility, remaining open to all others: "And to shew that I do not take Gravity for an essential Property of Bodies, I have added one Question concerning its Cause, chusing to propose it by way of a Question, because I am not yet satisfied about it for want of Experiments."[69]

Leibniz shared Newton's view of force as something more substantial than action at a distance through math or geometry alone. He characterized the purely kinematic view of Descartes as too one-sided for a complete understanding of natural phenomena, arguing that explaining the essence of matter requires a dynamic principle beyond the reach of extension and motion alone: "To give a foretaste of my conceptions, it is sufficient for me to explain that the notion of force or virtue which the Germans call 'Krafft,' the French 'la force,' and for the exposition of which I have designed a special science of dynamics, adds much to clear understanding of the concept of substance. For active force differs from the concept of bare power familiar to the Scholastics, in that this potentiality of faculty of the schools is nothing but a possibility ready to act, which nevertheless needs an external excitation or stimulus, as it were, in order to pass into action. But active force contains a certain activity or entelechy and is midway between the faculty of acting and the action itself; it includes effort and thus passes into operation by itself, without any auxiliary, but with only the removal of impediments."

Jammer writes, "For Leibniz, a moving body is different from a body at rest. Its motion is not merely a successive occupancy of different places in space, it is a state of motion at each separate moment. This state of continual change of place involves some effort. As, however, the very principle of inertia excludes an external influence for the continuance of this motion with constant speed, this effort must be the outcome of an activity inherent in the moving body. Moreover, inertia as the principle of resistance, to be overcome by the moving forces, must be of the same category as these, that is, it must be a force . . . In his monadology, Leibniz . . . relegates resistance or the force of inertia to the so-called prima materia, the final substratum of infinitely divisible matter. Furthermore, he even accounts for the extension of matter by this principle of resistance . . . Finally, since space is extended but obviously not impenetrable, impenetrability cannot be reduced to extension, and consequently also not to the inherent force of inertia. Primary matter, therefore, in Leibniz view, has two qualities: dynamical inertia . . . and an impenetrability (or, as Leibniz prefers to call it antitypia)."[70] Leibniz went on to employ his concepts of force to develop the mathematics of his dynamics, from which emerged the enormously important concept of energy, vis viva—the efficacy of force exhibited by a moving body.

Leibniz's ideas found currency with Immanuel Kant. In his Metaphysical foundations of natural sciences, Kant attempts to account for natural phenomena through the combinations of two primary forces, attraction and repulsion. He goes on to define the expansive force or elasticity of matter as the force of the extended by virtue of the repulsion of all its parts—a conception this author considers to be an overlooked and exceptionally efficient ontological integration. Upon this reasoning "Kant declares that by indefinitely increasing compressive forces opposing the forces of extension . . . matter can be compressed to infinity, but never penetrated. Further he shows that the very existence of matter requires a force of attraction as its second essential fundamental force. Otherwise, owing to the repulsive force, if acting without opposition, matter would dissipate itself to infinity and no assignable quantity of matter would be met with in any assignable space. These are the only two forces that can be conceived, claims Kant."[71]

It may be somewhat surprising to the reader to realize that Newton's, Leibniz's and Kant's philosophically rich opposition to the notion of force as action-at-a-distance did not attain great influence. Criticisms came forth along a line of reasoning stated by James Clerk Maxwell, "It is probable that many qualities of bodies might be explained on this supposition, but no arrangement of centers of force, however complicated, could account for the fact that a body requires a certain force to produce in it a certain change of motion, which fact we express by saying that the body has a certain measurable mass."[72] Presumably believing that space itself offers no resistance to impressed force, Maxwell was asking the question: if the innate nature of mass is simply force, why should it take force to move force?

Berkeley was among the most vocal opponents of an ontologically fundamental role for force: "Force, gravity, attraction and similar terms are convenient for purposes of reasoning and for computations of motion and of moving bodies, but not for the understanding of the nature of motion itself." He compared the employment of concepts of force in explanation of acceleration to the employment of 'epicycles' in explanation of planetary motions. In commentary that appears to mix Berkeley's view with his own opinion, Jammer writes, "All that natural science can supply is an account of the relations among symbols or signs; but the sign should not be confounded with the vera causa, the real cause of the phenomena. For real causes are active causes, they are productive, they make things occur. And real causes, as we have seen, are not the subject of physical science. The law of attraction, therefore, is to be regarded as a law of motion, and this only as a rule . . . 'observed in the productions of natural effects, the efficient and final causes whereof are not of mechanical consideration.'

"When science describes the motions given in direct experience and states them as simply as possible in rules or formulas, it may be useful to refer to 'force' or 'action' as an element in such formulas or equations, but then 'we are not able to distinguish the action of a body from its motion.' Science has no right to assume that bodies possess or exert real forces which enable them to act on each other. A close analysis of what we mean by body shows us that it contains nothing of this kind. If bodies are divested of all their qualities, as extension, impenetrability, and figure, all of which are passive qualities, nothing remains that can be interpreted as gravity or any other kind of force. 'Those who assert that active force, action, and the principle of motion are really in the bodies, maintain a doctrine that is based upon no experience, and support it by obscure and general terms, and do not themselves understand what they wish to say.' . . . Hume's thesis that the origin of the concept of causality is found in habit reduced causality to a mere association of perceptions. In fact, says Hume, we have 'no other notion of cause and effect but that of certain objects which have been conjoined together habitually in past experience,' and the possibility of understanding the nature of an objective connection between cause and effect must be denied . . ."[73]

The notion of action at a distance had indeed become a basic concept for the classical edifice of theoretical mechanics. Jammer writes, "Laplace, in his Mecanique celeste, stated that the objective of his study is a reduction of all mechanical phenomena to forces acting at a distance. Lagrange's monumental work, Mecanique analytique, the highlight of classical mechanics, was written in the same spirit. The mechanics of action at a distance gained further support in the successful applications of the classical theories of electricity and magnetism, as expounded by Laplace, Poisson and systematized by Weber. Even capillary effects, contact phenomena par excellence, were subjected by Laplace and by Gauss to the principle of action at a distance."[74]

Another of many prominent thinkers to jump on this bandwagon, Hertz expresses his logic through the example of a stone tied to a string and whirled around in a circular path: "But now the third law [of Newton] requires an opposing force to the force exerted by the hand upon the stone. With regard to this opposing force the usual explanation is that the stone reacts upon the hand in consequence of centrifugal force, and that this centrifugal force is in fact exactly equal and opposite to that which we exert. Now is this mode of expression permissible? Is what we call centrifugal force anything else than the inertia of the stone? Can we, without destroying the clearness of our conceptions, take the effect of inertia twice into account—firstly as mass, secondly as force? In our laws of motion, force was a cause of motion, and was present before the motion. Can we, without confusing our ideas, suddenly begin to speak of forces which arise through motion, which are a consequence of motion? Can we behave as if we had already asserted anything about forces of this new kind in our laws, as if by calling them forces we could invest them with the properties of forces? These questions must clearly be answered in the negative."[75] Though today it is obvious that these questions can, in fact, be answered affirmatively, Hertz's statement makes it clear how ambiguity in uses of the term 'force' contributed to the then-still-progressing rejection of common sense notions of force from physics.

Jammer continues, "The empirical, antimetaphysical attitude in mechanics, [would culminate] toward the end of the nineteenth century in the attempts of Kirchhoff, Hertz, and Mach to eliminate the concept of force from science . . . theories of mechanics that divested themselves of the concepts of cause and force (and finally also of the concept of substance) and adopted the purely functional [geometro-mathematical] point of view, taking force as a derived concept devoid of all temporal, causal, or teleological implications . . . The chief exponent of this doctrine, which may be called deanthropomorphic, positivistic, or operational, was Ernst Mach . . . [In his conception of mechanics,] force, as well as mass, are thus reduced . . . to purely mathematical expressions relating certain measurements found in space and time."[76]

Jammer eulogizes any fundamental ontological quality of force thusly: "The conception of force as the primordial element of physical reality, advanced by Leibniz, Boscovich, Kant and their followers, was not very fruitful and productive for the advancement of theoretical physics. It was a construct that was not easily assimilable into a conceptual scheme of operational import, and apparently remains such even in our atomic age which succeeded in releasing 'the force of the atom.' As far as physics proper is concerned, every modern physicist, most certainly, would agree with Thomson and Tait in calling such dynamic doctrine an 'untenable theory.'"[77]

Mach went so far as to say, "I hope that the science of the future will discard the idea of cause and effect, as being formally obscure; and in my feeling that these ideas contain a strong tincture of fetishism, I am certainly not alone."[78]

Indeed, built upon this line of reasoning, and equipped with the mathematics of Riemann, one man's startling restructuring of the epistemological framework of physics would soon fulfill Mach's hope.

# Chapter 3
# The Rise of Relativity

"Under the influence of the ideas of
Faraday and Maxwell the notion developed that the
whole of physical reality could perhaps be represented as a
field whose components depend on four space-time parameters.
If the laws of this field are in general covariant, that is, are not
dependent on a particular choice of coordinate system, then
the introduction of an independent (absolute) space is no
longer necessary. That which constitutes the spatial character of
reality is then simply the four-dimensionality of the field."[79]

—Albert Einstein

Albert Einstein's contributions to modern physics are extensive and profound, and go well beyond the theory for which he is most remembered. The first of his contributions concerned the nature of light, yielding the concept of the 'photon', the hypothesized behavior of which would later form part of the underpinnings of the theory of quantum mechanics. His second contribution concerned definition of the zigzagging of atoms, a phenomenon called Brownian motion. His third was the theory of relativity, evolving in two primary stages titled the special and general theories of relativity, respectively.

Before diving into the story of the development of relativity theory, let us summarize the state of physics at the beginning of the 20<sup>th</sup> century. Concepts of space, time, force, and mass incorporated into classical mechanics had proven quite sufficient for classical mechanics. Interestingly however, mechanics did not seem to need any absolute notion of space or time to maintain its theoretical or empirical integrity. Electrodynamics was proving effective in describing the behaviors of all kinds of electric and magnetic phenomena, and its predictions also appeared to depend only upon relative motions of the systems with which it dealt. The structure of atoms was becoming apparent as a constellation of smaller electrically charged bodies. Theories of the origin of inertia and gravitation in terms of electromagnetism or various kinds of ether were rampant. Mathematics and geometry were experiencing a profound revolution of their own as the non-Euclidean coordinate systems of Riemann and others were enticing and evocative of new ways of thinking about physics, subjecting notions of force and mass to new metaphysical questions. Newton's notions of absolute space and time seemed to lack any operational utility whatsoever. Meanwhile, the successes of physics were producing all sorts of amazing technological innovations.

Yet for all the progress made in physics up to the end of the 19th century, yielding numerous remarkably insightful principles that today underlie all technologies of human civilization, physicists did not share a driving consensus on basic, profound questions: why do forces move bodies the way they do? What, in truth, are 'space' and 'time'? What is 'motion'? What is 'mass'? What is 'force'? Indeed, what is a 'body'? While scientists had become equipped to measure the dimensions, motions and relations of things with remarkable precision, the ontology of physics remained, for the most part, ambiguous. In short, an air of pregnant anticipation was wafting through the minds of many physicists.

In this context, during the early years of the 20th century, a new theory of physics emerged that would thoroughly challenge prior assumptions concerning the fundamental metaphysics of science.

Einstein was deeply convinced by the empirical relativity of Newton's mechanics, Faraday's observations concerning electrodynamics and induction, and the philosophy of Mach that all motions are, in fact, strictly relative. He believed that there could be no such thing as an absolute frame of reference of any kind, and, like Mach, that notions of Earth moving through an ether were literally devoid of meaning. But there was a basic problem to be dealt with if this belief was, in fact, true: Maxwell's beautiful and effective theory of electrodynamics predicted two things: (1) that some kind of ether must exist through which light propagates, and (2) light travels at a certain constant speed, a speed that had been empirically verified.

One physicist at work on the problem of reconciling Maxwell's theory with the apparent relativity of electrodynamic phenomena was Heinrich Hertz. Hertz discovered a small change to Maxwell's equations needed to properly account for relative motion among bodies (in mathematical terms, replacing 'partial time derivatives' with 'total time derivatives'). It worked, and apparently resolved the paradox. But Hertz's tentative physical interpretation of his math—that the hypothesized ether moves along with a moving body—was later allegedly refuted, Hertz's math was thrown out with his physical interpretation, and thus his proposed correction was lost to 20th century physics.[80]

Into the vacuum left by the refutation of Hertz's physical interpretation of his math, Einstein stepped up with what was, by any measure, a bold assertion: both the absolute constancy of the speed of light and the relativity of physics should be retained in a completely new theory of relativity—a theory that would postulate no physical medium in space of any kind, moving with matter or not. The special theory of relativity was born.

In his landmark paper, On the Electrodynamics of Moving Bodies, Einstein stated, "It is known that Maxwell's electrodynamics—as usually understood at the present time—when applied to moving bodies, leads to asymmetries which do not appear to be inherent in the phenomena. Take, for example, the reciprocal electrodynamic action of a magnet and a conductor. The observable phenomena here depends only on the relative motion of the conductor and the magnet, whereas the customary view [a view that includes an ether] draws a distinction between the two cases in which either the one or the other of these bodies is in motion . . . Examples of this sort, together with the unsuccessful attempts to discover any motion of the earth relatively to the 'light medium' [ether], suggest that the phenomena of electrodynamics as well as of mechanics possess no properties corresponding to the idea of absolute rest. They suggest rather that . . . the same laws of electrodynamics and optics will be valid for all frames of reference for which the equations of mechanics hold good. We will raise this conjecture (the purport of which will hereafter be called the 'Principle of Relativity') to the status of a postulate, and also introduce another postulate . . . . Namely that light is always propagated in empty space with a definite [relative] velocity c which is independent of the state of motion of the emitting body."[81]

It is not clear whether Einstein was familiar with the results of the Michelson-Morley experiment at the time of his first statements on these questions. His words imply that he was, but we cannot be sure. In any case, his Principle of Relativity was consistent with their data—that indeed, the speed of light is measured to be constant relative to the motion of its observer.

Other physicists were convinced that the Michelson-Morley experiments should be interpreted differently, including at least one of the two men who conducted them (Michelson). Another was Dutch physicist Hendrick Lorentz. "He proposed that matter moving through the ether was shrunk by the ether in such a way that the measured speed of light, using shrunken rulers, was always the same, whatever the direction of propagation of the light ray or the motion of the ruler. In this way, light would appear to have the same speed in all directions, independent of the motion of the observer."[82] This notion was held in preference to Einstein's interpretation by many prominent physicists of the day, but for reasons we will visit shortly, and irrespective of whether there was merit in Miller's later alleged refutation of the original experimental result, Lorentz's alternative explanation of it failed to gain currency as the paradigmatic interpretation of the measured velocity of light.

The key to the rationalization of special relativity, in the words of Einstein, was that "My solution was really for the very concept of time, that is, that time is not absolutely defined but there is an inseparable connection between time and the signal [light] velocity."[83] Einstein believed that the speed of light, not the uniform flowing of time, should be considered the more fundamental concept in physics. Following from this solution, the common sense notion of the uniformity and simultaneity of time had to be abandoned. With his now-well-known mental picture of the train car containing a light clock and an observer, both being watched by an observer standing outside on the side of the train tracks, he demonstrated that his Principle of Relativity meant that time flows differently for different observers in motion relative to each other. In other words, events that appear to occur simultaneously to one observer would not necessarily appear to occur simultaneously to another observer. By leaving behind the common-sense assumption that the flow of time is uniform, he was able to promote the velocity of light into the status of a metaphysical absolute, and the implications were profound. In the words of the great mathematician Hermann Minkowski, "Henceforward space on its own and time on its own will decline into mere shadows, and only a kind of union between the two will preserve its independence."[84]

One of several implications of Einstein's theory was that any remaining vestiges of the force of inertia—Newton's vis inertiae—were stripped of any physical meaning. If light itself—which Einstein assumed traveled in the form of photons through a true void of nothingness—propagated at a speed strictly relative to every uniformly-moving observer, then uniform motion must in every sense be equated to the absence of motion, and, as Kant long before argued, who needs a force to explain the absence of motion? Thus, no recourse to a concept of force for the continuance of a body in motion was needed.

As significant as his work already appeared to be, Einstein was just getting started. His special theory of relativity dealt exclusively with the problem of light signals relative to uniform motion of a body. A few short years later he found the way to extend his principles to encompass the motions of bodies in gravity as well.

The first step in this process was what Einstein called his "happiest thought". Physicist Brian Greene writes, "In 1907, while pondering these issues at his desk in the patent office in Bern, Switzerland, Einstein had the central insight that, through fits and starts, would eventually lead him to a radically new theory of gravity—an approach that would . . . completely reformulate thinking about gravity and, of utmost importance, would do so in a manner fully consistent with special relativity."85 Einstein conceived the notion that weight and resistance to acceleration were indistinguishable. The mental picture is fairly easy to describe: imagine that you are in a sealed, opaque elevator floating in space. Now, imagine that something starts to pull the elevator along, accelerating it in the direction from your feet to your head. Though you cannot see outside of the walls to learn how you are moving, you will start to feel a force on your body that is proportional to the acceleration of the enclosure. Einstein asserted that this force is indistinguishable from the force of weight you feel while standing in gravity. Einstein reasoned that the mechanisms of inertial resistance and gravitational force must be one in the same, and he termed this the Principle of Equivalence.86 It was a profound and elegant insight, for it did what physics always seeks to do: explain more phenomena with fewer fundamental principles.

Having convinced himself that gravitational force and inertial resistance are indistinguishable, he asked himself a most logical question: might it also be the case that a body freely-falling in gravity—a body that is accelerating but which 'feels' no gravitational force—somehow is equivalent to a body in uniform-straight-line motion, in which it also 'feels' no inertial resistance? After reasoning on this question for several years, he found the answer he sought.

By employing Riemann's notions of non-Euclidean geometry, Einstein hypothesized that the dimensions of space, like that of time, need not be uniform. Instead, he suggested that matter 'warps' the actual dimensions of both space and time in such a way that the falling motion of a body in gravity is, in reality, the same kind of motion as uniform inertial motion, it's just that the geometry of 'space-time' is non-Euclidean near matter. In other words, the region surrounding a massive body like the Earth is not to be thought of as a field of force, but rather as a region in which the dimensions of space and time are invisibly warped by the matter of the Earth. Incorporating Lange's line of reasoning—that Newton's mechanics does not really depend upon absolute space, but rather only upon a coordinate system in which the laws of mechanics hold—Einstein reasoned that a freely falling (thus accelerating) body can be thought of as following a straight line—experiencing ordinary inertial motion—in the four-dimensional geometry of space-time. In simpler words, "mass grips space by telling it how to curve, and space grips mass by telling it how to move."87 The general theory of relativity was thus born.

Einstein published his new theory in 1915, and it was met with a mixture of awe and skepticism, for it called into question the entire apparatus of Newtonian physics. Indeed, it called into question the basic epistemological constructs of science, by asserting that our common sense of reality is not to be taken at face value. Understandably, most physicists chose to await empirical verification of the theory before they could accept such radical change.

They would not have to wait long, for on May 29, 1919, an expedition led by astronomer Sir Arthur Eddington tested a key prediction of Einstein's theory of general relativity: that rays of light are bent by the curved geometry of space-time as they pass near a massive body. On the island of Principe off the coast of West Africa, Eddington tracked the light of a star near the edge of the sun during a solar eclipse. On November 6, 1919, the results were announced to a joint meeting of the Royal Society and the Royal Astronomical Society: Einstein's prediction was confirmed. Greene writes, "It took little time for word of this success—a complete overturning of previous conceptions of space and time—to spread well beyond the confines of the physics community, making Einstein a celebrated figure worldwide. On November 5, 1919, the headline in the London Times read 'REVOLUTION IN SCIENCE—NEW THEORY OF THE UNIVERSE—NEWTONIAN IDEAS OVERTHROWN.' This was Einstein's moment of glory."88

In the decades since, the precision of Eddington's result has been called into serious question, but yet numerous experiments have been conducted that appear to establish the validity of the theories of special and general relativity. The motion of Mercury about the sun, the behavior of 'muon' particles, discovered in 1936, and other results have appeared to confirm relativity's predictions of time dilation, as did 1976 experiments comparing the timekeeping of atomic clocks in different conditions of motion. While there is also a considerable body of theoretical and experimental study alleging serious problems in both theories of relativity[89], the consensus view of modern physicists is that Einstein's theories are fully valid. Indeed, they are regarded by the mainstream physics community today as almost Biblical Truth.

Einstein's relativity theories transformed the metaphysical constructs of science in two ways that are obvious, and in at least two other ways that are more subtle, but no less significant. The obvious transformations involved the epistemological constructs of time and space in physics, and by extension, all other domains of science. Prior to relativity, time and space were viewed as metaphysically objective, ideal and uniform dimensions by which to measure subjective reality. Following relativity, the objectivity and uniformities of time and space were subordinated to local—dare I say subjective—measurements of clocks and rulers. Local clocks and rulers were appointed as the measures not of local clocks and rulers, but of the local dimensionality of time and space—of the very axes of local ontological and epistemological dimensionality.

Thus, in effect, Einstein's relativity incorporated into the theoretical apparatus of physics two postulates not commonly recognized for their full import: the only referent shared in all frames of measurement is a constant relative measure of the velocity of arriving and departing impulses, c, from which it must follow that frames of local measurement share no frame of reference in terms of space and time.

By implication, the less obvious transformations were physics' ontological concepts of the nature of force and mass. Prior to relativity, force (along with its expressions in energy, impulse, potential, etc.) and mass, while ontologically ambiguous, possessed undeniably fundamental ontological status. That is, prior to relativity, force and mass were conceived as the qualities that defined "thingness"—that which exists. Following relativity's elimination of gravitational force by replacement with the space-time dimensional curve, theoretical physics acquired a new mission: the elimination of the ontological status of force and mass altogether by explaining electromagnetic forces, nuclear forces, and atomic structure in terms of geometries of non-Euclidean space-time dimensions.

In effect, relativity took Newton's most basic epistemological frames of reference—dimensions of space and time—and warped them into physics' new ontology, destined to eliminate concepts of force and mass. At the same time, relativity took Newton's most basic ontological notions—force and mass—and mathematized them into physics' epistemological dimensionality, making physics' frames of reference absolutely relative.

To clarify these assertions, consider the following two diagrams. The first represents the dimensional epistemology through which the motions of mass were interpreted prior to Einstein:

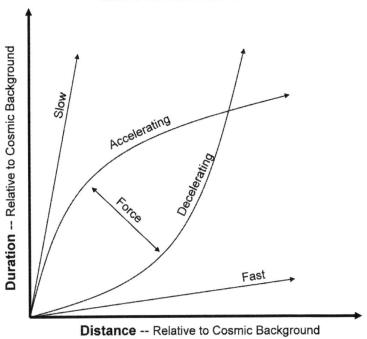

As we can readily see, the ability to discover a physical basis for patterns of motion arises through a shared epistemological frame of reference—axes of measurement—by which to characterize their differences. The second diagram reveals the effect of Einstein's interpretation of the relativity of physics on the epistemology of physical science. Notice that, in effect, canonical relativity has essentially 'morphed the axes of measurement' to conform to subjective paths of motion. In my opinion, this has served to conceal the ontological character of space, force, mass, particle, and perhaps even time, and it begs several basic questions: what, then, is the meaning of motion? In terms of what shall any kind of ontological dynamism of Nature be characterized?

Effect of Swapping Roles of Epistemology and Ontology in Physics

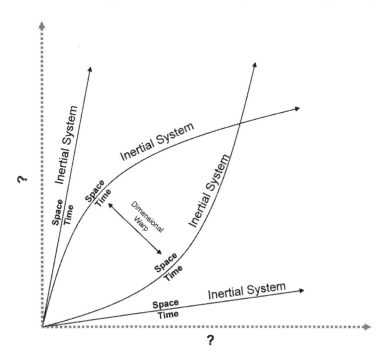

Though deep concerns about swapping epistemological and ontological constructs of physics were semi-consciously sensed by some scientists early in

the 20th century, most theoreticians over time became captivated by the sheer elegance of Einstein's union of space and time into a mathematically-describable Riemannian metric, and by the apparent economy of the resulting elimination of gravity as a force to reckon with. So seductive were the equations of these hypotheses that the man once declared "a lazy dog" by his mathematics tutor, a man appraised by his headmaster as unlikely ever to amount to anything, became in a few short years one of the most admired and influential scientists ever to walk the Earth.

# Chapter 4
# Welcome to Paradoxia

"Was this not revolution indeed?
Can one really assert that the time was ripe for a
revolution as radical as this?
It is probably the greatest mutation ever in the
history of human thought."90

—Jean Ullmo


It would not be long before the metaphysical implications of relativity began to work their way into the theoretical apparatus of science. A brief tour through theories of the 20th century reveals how empirical observations across entire orders of Nature's territories were theoretically interpreted in terms of Einstein's map.

## It All Began in a Really Big Bang

Among the first examples was witnessed in astronomical studies later to be organized as the discipline of cosmology. In the first few decades of the 20th century, astronomer Edwin Hubble was engaged in the task of measuring the distances of galaxies. The ruler he employed for this audacious effort was a period-luminosity relation observed in the light emitted from certain kinds of stars known as cepheid variables. The luminosity of these stars varies in regular patterns, such that they can be used somewhat like a 'standard candle', whose apparent brightness falls off with distance. Meanwhile, Vesto Slipher was measuring 'redshift' in the spectra of light from distant galaxies. He characterized this redshift as a Doppler effect in the light emitted by them, suggesting that they were moving with respect to the Milky Way. He found redshift in the light of each galaxy he measured, and the redshifts he observed indicated to him that all galaxies are moving away from the Milky Way.

In 1924, a number of scientists, including Carl Wirtz and Hubble, compared galactic distance measurements with Slipher's redshift measurements, and they were surprised to find a direct relationship. Subsequent collaboration between Hubble and Milton Humason confirmed the connection. The further away a galaxy is from us, the larger its redshift. This correlation became known as Hubble's law, and the implication seemed obvious: if the observed redshift is in fact a Doppler shift, then the galaxies are moving away from us, at greater speed with greater distance.

According to one modern textbook on the subject, "Hubble and Humason were careful to term the galaxies' rush away as 'apparent.' Taken at face value the velocities leave us fixed in the center of the universe. Having been thrust away from the center by Copernicus, astronomers felt somewhat uncomfortable at being repositioned there. Was the Milky Way Galaxy now enthroned as the center of the universe?"[91]

This fascinating empirical correlation was too much to resist for a young Belgian priest and aspiring relativist by the name of Georges-Henri Lemaitre. A student of Eddington, he heard a lecture by Hubble at Harvard Observatory, and learned of the redshift-distance relation. As a student of Einstein's theories, he was one of the then-few physicists capable of applying relativity to other questions of science, and that he did. Using the ontological and epistemological freedom Einstein granted theoretical physicists to manipulate the dimensions of space and time by equation, he put forth the hypothesis of an expanding universe in 1927. The same notion was independently, and previously, developed by the young Russian scientist Alexander Friedmann. In applying their theories to astronomical observations, they reasoned that we see galaxies flying away from us because the dimensionality of space itself is expanding. [92]

To better understand this assertion, an analogy is useful: imagine a community of ants scattered across a rubber sheet. If the sheet is stretched in all directions, each ant will see the others moving away from itself—the ants aren't crawling away from each other, they're being pulled apart by their underlying 'metric'. Before long, the obvious implication was reached: if the Universe was expanding, then it must have had its origin in some kind of point—a singularity from which all things in the Cosmos somehow sprung into being. This theory was not well received when first introduced. The only experimental evidence in its favor was circumstantial—the redshift of galaxies.

But the theory of an expanding Universe managed to survive until two important factors catapulted the hypothesis into the status of truth: good public relations, and an additional element of circumstantial evidence.

Eric Lerner writes, "To one of the Manhattan Project scientists, George Gamow, the detonation of an A-bomb constituted an analogy for the origin of the universe: if an A-bomb can, in a hundred-millionth of a second, create elements still detected in the desert years later, why can't a universal explosion lasting a few seconds have produced the elements we see today, billions of years later? . . . Unlike Lemaitre, Gamow had a tremendous flair for publicizing and popularizing his own theories, a flair that, within a few years, would establish his element theory . . . as the dominant cosmology."[93] Renowned astronomer Fred Hoyle attacked Gamow's theory, characterizing the belief that the Universe began in a "big bang" as fantastically illogical. The appellation stuck, however, and soon lost its derogatory character.

And then more circumstantial evidence emerged. Big bang cosmologists had predicted that if the Universe began in a potent explosion, there must be radiation left over, still traveling throughout space. In the mid-1960s, Arno Penzias and Robert Wilson, researchers at Bell Labs, observed a 'background' of microwave radiation arriving from all directions in space. While other processes in Nature might account for this radiation, no other prediction had the currency and convenience of the big bang theory at the time.

Alternative explanations of evolution of matter in the Universe—so-called steady-state theories, which vied for dominance with the big bang theory for several decades—eventually fell victim to these factors. In recent decades, observations—some of which we will examine later—have emerged suggesting that the big bang theory is utterly false, but they have been largely ignored by the community of science and by mainstream media.

Thus, employing the metaphysics of a theory built upon the principle that frames of local measurement share no frame of reference in terms of space and time—a stunning philosophical license blessed as Postulate in the epistemology of Einstein's interpretation of relativity—most human beings today are taught to believe that the Universe was created in a giant blast 15 billion years ago or so, from a tiny explosive seed which physicist Stephen Hawking suggests, in a nutshell, we refer to as a 'pea instanton'.

## Particle Soup

Einstein's interpretation of the relativity of physics and big bang's recipe for the historical creation of matter in a giant explosion have yielded a domain of theory called particle physics—a domain which might well be characterized as a branch of mathematics with varying correspondence to natural philosophy, depending upon the methods of its student.

The de facto abolition by relativity theory of all conceptions of 'spacetime' as a physical medium invited speculation that every appearance of 'force' in Nature may be regarded as a 'particle'. For if space is indeed nothingness in which special relativity's postulate of motion—or rather non-motion—rules, then what other kind of entity may cross its void than a discrete entity deserving its own special relativistic inertial frame? Upon this logic, particle physics leapt far beyond Thompson's humble concept of fundamental electrodynamic bodies—he called them electrons—and whatever kind of traveling density gradients—let's call them photons—that might otherwise have been ascribed to electrons' disturbances of the medium producing and/or arising from the infinite reach of their own superposed influences.

Timothy Ferris writes, " . . . the results of physics research are less finished than popular accounts usually suggest. The standard model of particle physics, though well worth boasting about, remains very much a work in progress. It makes accurate predictions, but to do so requires plugging in a couple of dozen parameters whose values are derived from experiment without benefit of any compelling underlying knowledge of why each should be what it is instead of some other number. Nor do the theories that compose the standard model dovetail into one another with any great logical inevitability. Three of the four fundamental forces of nature (interactions, in the jargon) are accounted for by quantum field theories that incorporate the special theory of relativity. But the fourth force, gravitation, remains the province of general relativity, which is excluded from the quantum theories and which, moreover, views the world in a very different way than the quantum theories do. Needed—or, at least, desired—is a single theory that would embrace all four forces, and in doing so lay bare the reasons that the experimentally determined values are what they are . . . Wise hunters stalk the ultimate theory by searching for signs of symmetry. The laws of nature are all expressions of symmetry, and all physics is in some sense a search for symmetry."[94]

In the words of Lerner, "In an effort to introduce some order and symmetry . . . particle theorists developed the 'standard model' . . . The model assumes that all the forces of nature are quantized—that each force is carried by [discrete] particles. These force-carrying particles are exchanged [among] other particles, thereby generating force. The model hypothesizes that mesons and baryons are made up of six different types of particles called quarks. And then there are the six leptons: electrons, muons, tauons, and three neutrinos. A photon carries electromagnetic force, the W and Z particles carry the weak force, and no less than eight gluons carry the strong force—hence their name, because they glue one particle on another."[95]

The theory of matter creation in big bang cosmology directly intersects this domain of fundamental physics. The following post-big-bang timeline has been hypothesized by cosmologists and particle physics theorists: before $10^{-43}$ seconds after the big bang, there was perfect symmetry; at $10^{-43}$

seconds after the big bang, gravity separated into its own force; at $10^{-35}$ seconds after the big bang, the strong force separated into its own force, as quarks and leptons became distinct through the inflation of space itself; at $10^{-12}$ seconds after the big bang, the weak force separated; protons formed at $10^{-5}$ seconds after the main event; at 250 seconds, helium formed; 700,000 years later, atoms formed and a cosmic background radiation remained as the energetic fog cleared; one billion years or so after the big bang, stars, galaxies, and superclusters of galaxies emerged; and some 12-18 billion years or so after the Universe was born, humans emerged to contemplate all of this.[96]

Most competent physicists do not really believe that the panoply of particles described in the standard model share no fundamental ontological character, but it seems clear that the overwhelming tendency to focus on the labeled, mapped entities may be obscuring the crucial and common nature of the underlying territory. This is not intended to imply that there is no empirical data to support the constructs of particle physics. Decades of precision data from particle accelerators indeed speaks volumes about the behaviors of Nature. Rather, my point is that Nature's observable interactions, correlated by theoreticians to the 'particles' of the standard model, have been interpreted within the space-time epistemology of special relativity—yielding an ontological conception of reality absent the notion of a physical medium pervading, and probably comprising, all forms of matter.

Brilliant dissident physicist Thomas Phipps airs a view held by a large fraction of empirically-minded physicists whose voices are routinely suppressed by the academic peer review of institutional physics today: "Let this single excerpt from the article 'Quantum gravity: the last frontier,' by Gary Gibbons, Cambridge University lecturer on the Department of Applied Mathematics and Theoretical Physics—a department title that says it all, right there, concerning the victory of the [mathematical] barbarians in the political power struggle that has sent physics to the back of its own bus—(New Scientist, 31 October 1985, page 45) speak for all the rest:

"'Supergravity theories are supersymmetric because, in addition to the graviton, they introduce one or more 'gravitinos' which are fermions with 3/2 units of angular momentum. The rule is that there must be just as many fermions as there are bosons. Because the theory allows no more than eight gravitons, this rule forces us to introduce 28 spin-1 particles and 35 spin-0 particles, together with 64 spin-1/2 particles. All 128 particles are on the same footing in this theory, which is called 'N-8 extended supergravity' or 'N-8' for short. By introducing all these extra particles and putting fermions and bosons together, theorists originally hoped that the quantum fluctuations would be finite. We do not yet know for sure, but physicists now generally feel that this is unlikely. In any case, N-8 has other problems. The theory predicts that particles can interact by non-gravitational forces, such as the electromagnetic force, only if spacetime is very highly curved and the Universe consequently very small (about $10^{-33}$ centimeters).'"[97]

Right.

## Dark Energies

When Einstein was formulating his equations, he was concerned about one implication of his theory of gravity: it suggested that over great stretches of time, all the matter in the Universe ought to clump together from its mutual gravitation. A believer in the notion that the Universe must be 'static'—generally unchanging over cosmic time—he decided to include a term called the 'cosmological constant' which postulated an omnipresent repulsive force throughout space, balancing the tendency of matter to collapse together.

Einstein would later characterize the addition of this term as his greatest blunder, after other theoreticians, armed with the theory that the dimensions of Universe themselves were expanding following a big bang, demonstrated that an expanding Universe held no need for such a cosmological repulsive force—rather, the expansion process itself prevents matter from collapsing. In the modern words of Hawking, "This was one of the missed opportunities of theoretical physics. If Einstein had stuck with his original equations, he could have predicted that the universe must be either expanding or contracting."[98]

Yet in the past few years, new observations of the 'expanding' Universe have taken an even more exotic turn, motivating scientists to reconsider whether Einstein was right to postulate a universal repulsive force. Astronomers studying the light arriving at Earth from distant supernovas have noticed that their luminosities appear different than they ought to appear if the Universe has been expanding at a uniform rate since the big bang. As popularly reported, "[The new data] shows that the expanding of the universe is really speeding up and not slowing down as conventional astronomers had thought for 70 years,' University of Chicago astronomer Michael Turner told reporters . . . The new stellar explosion has helped astronomers understand how the universe expands, 'much the same way a parent follows a child's growth spurts by marking a doorway,' said Hubble scientist Adam Riess, lead researcher in the new study."[99]

Fueling this reconsideration are additional observations indicating that the big bang happened too recently to account for the calculated ages of galaxies and stars. This discrepancy has motivated theoreticians to postulate that, if some hitherto unrecognized cosmic force is increasing the pace of the Universe's expansion, we might have been assuming that the big bang happened more recently than it actually did, since expansion would have been slower in the past, and thus galaxies and stars would have had the time to form into the state they presently occupy. Davies writes, "In October 1994, sensational new results from the Hubble Space Telescope were announced, [from which some] commentators inferred an age of the universe of only eight billion years. The discrepancy with stellar ages is now glaring, and the cosmic age paradox has forced its way back onto the scientific agenda. While some cosmologists began questioning the standard big bang scenario, Hubble team member Barry Madore suggested that the cosmological term could stage a comeback. He told the Boston Globe, 'Einstein had the answer in his hands when he first formulated general relativity.'"[100]

Thus, the prevailing motion picture of cosmology is now this: before the big bang, we know not what existed, for space and time themselves did not. Then the Cosmos was begat in a giant explosion of a 'pea instanton' in which the dimensions of space and time experienced a dramatic period of 'inflation', the fundamental forces 'separated', all the fundamental particles emerged out of the energetic soup, the Universal expansion then began to slow down until a mysterious repulsive force overtook the attraction of gravity, and now the force of this 'dark energy' is causing the expansion of the Universe to accelerate.

Rube Goldberg would be proud.

## Imponderably Black Holes

In late 1915, Karl Schwarzchild, director of the Potsdam Observatory, was among the first scientists to employ Einstein's brand-new field equations. Schwarzchild presented a mathematical analysis of the gravitational field outside a uniform ball of matter. It correctly yielded Newton's law of gravity at large distances. But something odd about the solution puzzled Einstein throughout his life: at a certain critical radius, later known as the Schwarzchild radius, time would become infinitely warped. Einstein appraised this notion to be highly unphysical, but it didn't seem a serious problem, because, in considering the example of the space-time warp around our star, the sun would have to be compressed to a size smaller than Earth for such a singularity to develop. Such an idea seemed far-fetched, to say the least.[101]

Then in 1930, nineteen-year-old student Subramanyan Chandrasekhar was exploring equations describing a mysterious class of stars called white dwarfs. As Davies writes, "To his bafflement, Chandrasekhar found his sums predicting an outlandish result. If the white dwarf had a mass of more than 1.4 suns, then, according to the calculation, it could not remain stable, and would collapse further, without any apparent limit." From such a collapsed body, nothing—not even light—could escape the gravitational well. Einstein, like most of his peers, remained highly skeptical. He said that an infinite timewarp "does not appear in nature for the reason that matter cannot be concentrated arbitrarily." Eddington similarly rejected the notion, characterizing as absurd "a magic circle which no measurement can bring us inside."[102]

The debate over the possibility of gravitational singularities continued through subsequent years, but shifted into high gear when a high-altitude rocket equipped with a primitive X-ray detector registered a strong signal in the constellation of Cygnus. Davies writes, "It was dubbed Cygnus X1, and ten years later it became the first candidate object for a possible black hole formed by stellar collapse . . . The early and mid-1960s also marked important theoretical advances. British mathematician Roger Penrose developed new and much slicker geometrical techniques for studying Schwarzchild's spacetime, event horizons, collapsing stars, singularities and related aspects of the general theory of relativity. These new methods were to prove a boon for physicists struggling to come to terms with the bewildering properties of black holes.

"Finally came the discovery of pulsars (neutron stars) in 1967. By this stage gravitational collapse, supernova explosions, frozen stars and infinite timewarps were firmly on the astrophysicists' agenda. In late 1967, a conference on pulsars was held in New York, and Wheeler referred to the possibility that continued collapse would produce a 'black hole' in space. The name had finally entered the English language."[103]

Research on the subject of black holes blossomed worldwide throughout the 1970s and 80s, and today black holes are a staple of every textbook on astrophysics. No empirical data exists that conclusively establishes the existence of such objects with the nature defined in general relativistic theory, but the properties of numerous massive X-ray sources have convinced most astrophysicists of their reality. It is now generally thought that most—perhaps all—galaxies, including our very own Milky Way, harbor at least one region at their centers where the dimensions of space and time themselves, along with the laws of physics, contort into imponderability.

## Dark Matters

In recent decades, astronomers have discovered a striking discrepancy between the predictions of physics' theories of gravitation and the motions of stars within galaxies. According to Stephen Hawking, "Various cosmological observations strongly suggest that there should be much more matter in our galaxy and other galaxies than we see. The most convincing of these observations is that stars on the outskirts of spiral galaxies like our own Milky Way orbit far too fast to be held in their orbits only by the gravitational attraction of all the stars that we observe . . . This discrepancy indicates that there should be much more matter in the outer parts of the spiral galaxies . . .

"Cosmologists now believe that . . . the central parts of spiral galaxies consist largely of ordinary matter that we cannot see directly . . . Before the 1980s it was usually assumed that this dark matter was ordinary matter comprised of protons, neutrons, and electrons in some not readily detectable form, perhaps gas clouds, or MACHOs—'massive compact halo objects' like white dwarfs or neutron stars, or even black holes. However, recent study of the formation of galaxies has led cosmologists to believe that a significant fraction of the dark matter must be in a different form from ordinary matter. Perhaps it arises from the masses of very light elementary particles such as axions or neutrinos. It may even consist of more exotic species of particles, such as WIMPs—'weakly interacting massive particles'—that are predicted by modern theories of elementary particles but have not yet been detected experimentally."[104]

The resonant echo of the mainstream media promotes this hypothesis ever closer to certainty. As but one of hundreds of examples, MSNBC reports: "Dark matter, a mysterious form of matter unlike familiar atoms and subatomic particles, came into being almost immediately after the Big Bang, long before ordinary matter formed. Dark matter acted as gravitational 'seeds' for the density variations to grow. Because these areas of higher density had stronger gravity than their surroundings, they attracted more matter, and eventually grew to become the seeds of galaxies. (Today, galaxies are surrounded by dark matter halos, which are thought to be 10 times larger and more massive than the galaxies' visible portions.)"[105]

A growing community of theoreticians is pursuing this idea actively today, asserting some remarkably exotic notions. Space.com reports, "If the concept of dark matter gives you a bit of a headache, hold on to your Advil. Theorists attempting to explain some of the 'missing mass' in the universe now say there may be entire galaxies that are dark. The new idea, proposed by Neil

Trentham of the University of Cambridge, along with colleagues Ole Moller and Enrico Ramirez-Ruizof, suggests that for every normal, star-filled galaxy, there may be 100 that contain nothing, or at least nothing that we understand. This so-called dark matter—which may or may not be composed of the same particles that form you, me and the fence post—would in this scenario have created numerous low-mass mini-galaxies that lurk in what is otherwise known as intergalactic space."[106]

Several substantially less exotic alternatives have been advanced to explain the discrepancy between prevailing theories of gravity and the motions of cosmic bodies, such as those derived from the cosmic electrodynamics postulated by Nobel laureate Hannes Alfven.[107] But most often, they have confronted a wall of exclusion from broad publication and well-financed study, by virtue of an almost religious conviction among physicists that Newton's law of gravity and Einstein's dramatic evolution thereof can admit no future evolution. In his biography of Einstein, physicist Abraham Pais writes, "Einstein's discovery appealed to deep mythic themes. A new man appears abruptly, the suddenly famous Dr. Einstein. He carries a message of a new order in the universe . . . His mathematical language is sacred, . . . the fourth dimension, light has weight, space is warped. He fulfills two profound needs in man, the need to know and the need not to know but to believe . . . he represents order and power. He became the divine man of the twentieth century."[108] Through the lens of peer review today, one who proposes significant corrections to the force laws governing cosmic-scale motions is doing violence to a Sacred Law.

Thus, ironically at odds with the positivist approach to exploring a mystery of mass in motion, the mainstream scientific community and the educated public believes that as much as 90% of the matter in the Universe is comprised of unseeable stuff. While this may or may not be true, it is now increasingly assumed that this dark matter is 'exotic'—a substance on the one hand unlike the ordinary matter of which we and everything else we see are formed, but on the other hand faithfully causing and following the warps of Einstein's dimensions. Astronomer Riccardo Scarpa of the European Southern Observatory suggests: "Dark matter is the craziest idea we've ever had in astronomy. It can appear when you need it, it can do what you like, be distributed in any way you like. It is the fairy tale of astronomy."[109]

## Shortcuts Through Space, Travels Through Time

The exotica afforded by the authority to manipulate the dimensions of space and time by equation is not limited to big bangs, for as physicist and prolific author Michio Kaku observes, " . . . after Einstein's death, so many solutions of Einstein's equations have been discovered that allow for time machines and wormholes that physicists are now taking them seriously."[110]

As fictionalized in Carl Sagan's book and Warner Brothers' hit movie, CONTACT, a wormhole is a two-sided warp in the dimensions of space, connected by a kind of dimensional tube conceivably allowing one to cross arbitrary cosmic distances—even venture into 'other universes'—without actually traversing the three-dimensional Universe we empirically observe.

Kaku writes, " . . . in 1963 mathematician Roy Kerr found the first realistic description of a rotating black hole. Instead of collapsing to a point, as in a stationary black hole, it collapsed to a rapidly rotating ring of neutrons. The fact that the black hole is spinning is crucial: the centrifugal force keeps the ring from collapsing into a point. Kerr proved that anyone falling through the ring would not die, as commonly thought, but would actually fall through the ring into another, parallel universe."[111]

Using similar mathematical constructs, it is possible to conceive of time travel—creating a loop-like warp in the dimension of time that would, in theory, allow you to venture off into history and the future. Stephen Hawking writes, "Time travel is possible in a region of spacetime in which there are time loops, paths that move at less than the speed of light but which nevertheless manage to come back to the place and time they started because of the warping of spacetime."[112]

Such renowned figures as Hawking and Kip Thorne, along with a large number of other less well-known theoreticians, now routinely publish papers and books on wormholes and time travel—generally received with appropriate awe by media and public audiences thirsting for visions that reach beyond the Dow Jones ticker.

## Why Only One Universe?

In the first several decades of the 20th century, observations were made indicating fundamental particles and electromagnetic radiation interact in discrete units, called quanta. It was determined that are certain lower limits on the magnitude of fundamental interactions among bodies, and the mapping and correlation of interactions at this level yielded the theory of quantum mechanics. In the realm of the microscopic behavior of charges and atoms, strange behaviors seem to take over our otherwise common sense picture of cause and effect. For example, we have been unable to find a way to experimentally observe simultaneously both the velocity and position of a moving electron. It appears that the more we resolve either velocity or position of the ever-moving electron, the less we can resolve the other measure of it. This behavior is called the Heisenberg Uncertainty Principle, named after the scientist who conceived the relation. Physicists have also observed strange simultaneous correlations of actions at this quantum level—where changing the state of one particle in one place apparently instantly results in an opposite change of a 'correlated' particle in another place, without a signal having time to traverse the distance between the two. Experiments seem to suggest that these correlated actions do not actually occur until an observation is made of one of the particles—yet another alleged violation of our common sense notions of how things work.

Since the time of Heisenberg, physicists have developed models which attempt to make some sense of these weirdnesses by hypothesizing that fundamental particles occupy several states of energy and motion at the same time. A system of particles in this 'superposition of states' 'chooses' a particular state when it is observed, according to a statistical probability function. But before the act of observation, it is commonly asserted, the particles are believed to be following all possible paths all the time.

In intervening decades, some physicists and cosmologists have latched onto this notion and speculated that the Universe we see represents just the one set of states and paths followed by fundamental particles that have experienced the one path of evolution we see. They assert that, if all fundamental particles are actually occupying many states and paths at once, then there are in fact 'many worlds' unseen to us, in which different events transpired than the ones of which we see record. For example, in your world, you decided to eat breakfast yesterday, just as you remember. But in another 'parallel world', you didn't eat breakfast, but instead were killed in a traffic accident on your way to work. These theorists suggest that many universes do, in fact, exist, and that their realities are described by the total probability functions of all fundamental particles in the Universe.

This line of reasoning has led to an extensive body of professional and popular literature on the alleged 'parallel universes' or 'multiverse' in which we really live, though we can sense only the one universe we sense.

Kaku writes, "According to this startling new picture, in the beginning was Nothing. No space. No time. No matter or energy. But there was the quantum principle, which states that there must be uncertainty, so even Nothing became unstable, and tiny particles of Something began to form.

"By analogy, think of boiling water, which is a purely quantum mechanical effect. The bubbles, which seem to come out of nowhere, suddenly expand and fill up the water. Similarly, in this picture Nothing begins to boil. Tiny bubbles began to form and then expand rapidly. Since each bubble represents an entire universe, we use the term 'multiverse' to describe the infinite ensemble of universes. According to this picture, our universe was one of these bubbles, and the expansion is called the Big Bang.

"At first, creating bubbles of Something in an ocean of Nothing might seem to violate the conservation of matter and energy. However, this is an illusion, because the matter-energy content of the universe is positive, but the gravitational energy is negative. In fact, the sum of the two is zero, so it takes no net energy to create a universe out of Nothing! . . .

"Cosmologist Stephen Hawking believes that our universe is perhaps the most likely of all these infinite universes. In his picture, we coexist with an infinite sea of other bubbles (which he calls baby universes), but our universe is special. It is the most stable, and its probability of existing is the largest. He believes that all these baby universes are connected to each other by an infinite network of thin wormholes. (In fact, by adding up the contribution of these wormholes, he can present arguments why our universe is so stable.) These wormholes are extremely small, so we do not have to worry about falling into one of them and finding ourselves in a parallel universe.

"Steven Weinberg finds the idea of the multiverse an appealing one: 'I find this an attractive picture and [it's] certainly worth thinking about very seriously. An important implication is that there wasn't a beginning; that there were increasingly large big bangs, so that the [multiverse] goes on forever—one doesn't have to grapple with the question of it before the bang. The [multiverse] has just been here all along. I find that a very satisfying picture.'"[113]

I wonder if Ptolemy would have been satisfied with this picture?

## Things as Strings

Another thread of physics is yielding a similar conception of the nature of Nature. Einstein's theories of relativity—asserting explanations for the large-scale motions of matter—are broadly recognized as incompatible with the theory of quantum mechanics—describing the small-scale workings of matter. The empirical fact that brings this conflict into focus is nicely described by Greene: "The notion of a smooth spatial geometry, the central principle of general relativity, is destroyed by the violent fluctuations [observed in] the quantum world on short distance scales. On ultramicroscopic scales, the central feature of quantum mechanics . . . is in direct conflict with the central feature of general relativity . . . Like a sharp rap on the wrist from an old-time schoolteacher, [this] is nature's way of telling us that we are doing something that is quite wrong."[114]

In attempting to reconcile this vexing problem, a new hypothesis has rocketed to the top of the physics charts in the past ten years: string theory. The genesis of string theory traces back to a theoretical physicist named Gabriele Veneziano. While studying the strong nuclear force—the force hypothesized to hold together charges within atoms—he discovered that a mathematical relation, formulated by Leonard Euler for other reasons two centuries earlier, seemed to explain numerous properties of strongly interacting charges. Then in 1970, Yoichiro Nambu, Holger Nielsen, and Leonard Susskind showed how this correspondence might be explained: if elementary particles were conceived as tiny, vibrating, 'one-dimensional strings', then their behavior could be described by Euler's math. If the strings were small enough, then from a distance they would still appear to be point particles.[115]

In the following few years, the notion of particles as strings lost ground when more refined experiments contradicted the predictions of string theory, and when theorists realized that other particles were predicted that are not observed. But in 1974, Schwarz and Scherk transformed the latter liability into an asset, by showing that the predicted particles had properties matching those of another particle hypothesized to explain gravitational force—the graviton. The graviton has never been detected to this day, but in 1984, Green and Schwarz published a paper that appeared to point a way in which string theory could reconcile other conflicts with experimental data. At the same time, they hypothesized how string theory might be able to describe not just the strong force and gravity, but the weak and electromagnetic forces as well. The response from the physics community was overwhelming. Over the subsequent three years, over a thousand research papers were published exploring various aspects of string theory.[116]

Brian Greene describes the Universe in terms of string theory thusly: " . . . according to string theory, the observed properties of each elementary particle arise because its internal string undergoes a particular resonant vibrational pattern. This perspective differs sharply from that espoused by physicists before the discovery of string theory; in the earlier perspective the differences among the fundamental particles were explained by saying that, in effect, each particle species was 'cut from a different fabric.' Although each particle was viewed as elementary, the kind of 'stuff' each embodied was thought to be different.

Electron 'stuff,' for example, had negative electric charge, while neutrino 'stuff' had no electric charge. String theory alters this picture radically by declaring that the 'stuff' of all matter and forces is the same. Each elementary particle is composed of a single string—that is, each particle is a single string—and all strings are absolutely identical. Differences between the particles arise because their respective strings undergo different resonant vibrational patterns."[117]

In order to reach these hypothetical feats of apparent integration, string theory demands an even stranger alteration of our conceptions of the Cosmos than Einstein's relativity. The most recent versions of string theory, integrated under an umbrella M-theory, demand that the three dimensional Universe we observe actually be comprised of many 'hidden' dimensions, which we cannot see because they are curled up in forms too small to be detected by experimental means—now and possibly ever. As Greene writes, " . . . to accomplish these feats, it turns out that string theory requires that the universe have extra space dimensions . . . For string theory to make sense, the universe should have nine spatial dimensions and one time dimension, for a total of ten dimensions . . . [though the original] calculation underlying [this conclusion turned] out to be approximate . . . [so] the approximate calculation actually [missed] one additional dimension."[118]

It gets more strange: M-theory appears to predict the attributes of known elementary particles, but "a fundamental property of [the theory] is that it is highly symmetric, incorporating not only intuitive symmetry principles but respecting, as well, the maximal mathematical extension of these principles, supersymmetry. This means . . . that patterns of string vibrations come in pairs—superpartner pairs—differing from each other by half a unit of spin . . ."[119]

And it gets stranger still. In order to account for transformations of objects hypothesized in M-theory, an even more exotic notion is required: that the dimensionality of space can 'tear'. The Riemannian geometry underlying Einstein's general relativity does not allow for such notions, but M-theorists are counting on new mathematical constructs to describe the 'tearing' of space:

"In 1987, Shing-Tung Yau and his student Gang Tian . . . made an interesting mathematical observation. They found, using a well-known mathematical procedure, that certain . . . shapes could be transformed into others by puncturing their surface and then sewing up the resulting hole according to a precise mathematical pattern . . . Mathematicians call this sequence of manipulations a flop-transition . . . By late 1991 . . . at least a few string theorists had a strong feeling that the fabric of space can tear . . ."[120]

You may well wonder whether there might be catastrophic consequences to the 'tearing of space'. But Brian Greene, one of the pioneers of related concepts in string theory, explains, "The resolution to [this issue] relies on a central feature of quantum mechanics . . . [In] Feynman's formulation of quantum mechanics, an object, be it a particle or a string, travels from one location to another by 'sniffing out' all possible trajectories. The resulting motion that is observed is a combination of all possibilities, with the relative contributions of each possible trajectory precisely determined by the mathematics of quantum mechanics . . . It is these contributions that Witten showed precisely cancel out the cosmic calamity that the tear might otherwise create. In January 1993, [we] released our papers simultaneously . . . The long-standing question of whether the fabric of space can tear had been settled quantitatively by string theory."[121]

But even among mainstream theoreticians, M-theory is by no means accepted as truth. In fact, many scientists today view the foundations of this theory as thoroughly indefensible. In the mid-1980s, Nobel prize winner Sheldon Glashow and Paul Ginsparg wrote: "In lieu of the traditional confrontation between theory and experiment, superstring theorists pursue inner harmony, where elegance, uniqueness and beauty define truth. The theory depends for its existence upon magical coincidences, miraculous cancellations and relations among seemingly unrelated (and possibly undiscovered) fields of mathematics. Are these properties reasons to accept the reality of superstrings? Do mathematics and aesthetics supplant and transcend mere experiment?" Glashow has tempered his criticism in recent years, but not for any reason of empirical validation.[122]

More concerning, the following quote from theorist David Gross forecasts the justification to be used to support decades of theoretical research in string theory in the absence of a shred of experimental evidence: "It used to be that as we were climbing the mountain of nature the experimentalists would lead the way. We lazy theorists would lag behind. Every once in a while they would kick down an experimental stone which would bounce off our heads. Eventually we would get the idea and we would follow the path that was broken by the experimentalists. Once we joined our friends we would explain to them what the view was and how they got there. That was the old and easy way (at least for theorists) to climb the mountain. We all long for the return of those days. But now we theorists might have to take the lead. This is a much more lonely enterprise."[123]

The concepts of string theory may strike the empirically-minded scientist as, we might conservatively say, 'loose' physics, especially since there exists no experimental test by which M-theory has been or presently can be empirically supported.

But for experimentally-challenged theoreticians today, what's a class of elementary particles, an extra six or seven dimensions, and minor tears in the dimensions of space among peers?

~

If the above stew of theoretical imaginations—big bangs, dark energy, dark matter, particle soups, black holes, wormholes, time travel, parallel universes, and string theories—leaves your stomach a bit unsettled, you are not alone. Unfortunately, the words of the brilliant dissident physicist Thomas Phipps likely sum things up pretty well: "Physics, a microcosm of the larger society, has suffered its own invasion by barbarians; viz., hordes of pseudo mathematicians. It is typical of barbarians through the ages that they have no feeling, respect, nor understanding for the cultures they invade and destroy. The mathematical barbarians who have destroyed theoretical physics have yet to grasp the significance of the method of successive approximations, lack all conception of logical economy, and scorn operational definitions. Devoid of feeling for the physical, they acknowledge no limitations on imagination and obliterate all distinction between science and science fiction. They have imposed their own culture, straight from mathematics, which involves the Einsteinian instant leap into the bosom of final exact Truth . . ."[124]

All of the exotic modern theories reviewed above, and others, are directly or indirectly founded upon the following postulated interpretation of relativity catapulted by Einstein into the theoretical apparatus of modern physics: the only referent shared in all frames of measurement is a constant relative measure of departing and arriving impulses, c. Such a postulate seems an awfully slippery curve upon which to rest the dimensional epistemology of science and the ontology of Nature, particularly when there is, in fact, a frame of reference shared by all empirical frames of measurement: the observable Universe.

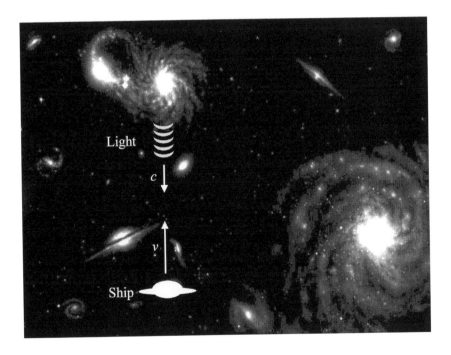

Measured within this empirical, shared frame of reference, the velocity of impulses—the speed of light—relative to moving bodies is most certainly not a constant. A simple diagram expresses this fact:

In the above diagram, a ship is traveling towards a galaxy at velocity v. Light from the galaxy is traveling in all directions at velocity c. It is apparently true that the ship will, in its moving frame of reference and using Michelson-Morley type apparatus, measure the incoming light waves as having velocity c regardless of the motion of the ship or the galaxy. But if instead we employ the empirical, observable Universe as our frame of reference (the closest approximation for which is likely the topography of the cosmic background radiation), it is quite clear that the relative velocity between the ship and the incoming light waves is v + c. Relativists argue that such a frame of reference is impermissible—or at best irrelevant—to employ for such a measure. I—and a significant minority in the physics community with roots stretching back at least to the early 20th century—thoroughly disagree, as shall be explained below.

In the words of Robert P. Kirshner, "Although the universe is under no obligation to make sense, students in pursuit of the Ph.D. are." I fear that 20th century theoretical physics slipped and fell into the warped dimensions of an absolutely relative epistemology, incorporating a debilitating subjectivity into the theoretical interpretation of empirical measurements, thus in the process mathematically concealing the objective ontological dynamis through which physics matter.

# Chapter 5
## It's OK to be Empirical

*"A Hair perhaps divides the False and True*
*Yes; and a single Alif were the clue—*
*Could you but find it—"[125]*

—Fitzgerald

Setting aside the exotica of modern theoretical physics outlined above—the philosophical foundations for which ought challenge any rational mind—the docket of empirical evidence supporting the case that theoretical physics is seriously off course fairly bulges at the seams. Consider just six brief examples, among many more:

1. Research in the fields of quantum mechanics, nuclear physics, and chemistry has revealed that every charge in every atom in every molecule across the Universe continuously experiences a 'jostling' motion, even at the temperature of absolute zero. Whether we choose to regard this fact as Principle by reference to Heisenberg's assertion of intrinsic uncertainty in physical processes, or instead to a hitherto unmapped motion of elementary bodies in a force-giving medium, it is plainly clear that no fundamental body anywhere in the Universe follows the path of motion which forms the basis for the definition of special relativity's inertial frame: uniform straight-line motion. Consider a trivial example: a billiard ball thrown in space, distant from any gravitating body. One intuitively would assume that its motion complies with the dictates of Newton's first law of motion and Einstein's special relativity—namely, the ball executes uniform

2. straight-line motion. But is this, in fact, true? Is there any single particle within the billiard ball executing straight-line motion? Is it not the case rather that every single elementary particle within the ball is experiencing continuous acceleration? Doesn't every particle in the ball, in fact, move along a path that is, at best, unaddressed by postulates of special relativity? Is the response from the relativist—that the average of particle motion is linear—an acceptable escape from the implication of the answers to these questions?

3. Experiments have conclusively confirmed that some kind of energetic faculty pervades the cosmic medium within which all matter exists. Known as the Casimir effect, a strange attraction is exhibited by closely-spaced quartz or metal plates, explicable only if the medium is itself in some sense dynamic. Numerous other experiments confirm that the 'vacuum of space' is, in fact, an energetic medium. Thus, it cannot be said that space is a void—an essential philosophical assumption incorporated into the train of logic that led to the epistemological and ontological framework postulated in Einstein's interpretation of the relatively of physics

4. The big bang theory of creation and dimensional expansion of the Universe is thoroughly incompatible with recently revealed empirical observations of extragalactic redshifts. As renowned radio astronomer Halton Arp and others have demonstrated, high redshift objects—known as quasars—are now observed emerging from the centers of low redshift galaxies, suggesting that galaxies give birth to galaxies, and that quasars are, in fact, young galaxies in formation.126 Arp writes, "I believe the observational evidence has become overwhelming, and the Big Bang has in reality been toppled. There is now a need to communicate the new observations, the connections between objects and the new insights into the workings of the universe—all the primary obligations of academic science . . ."[127] I suggest that Arp's data may also rationalize four other empirical mysteries. First, I suggest that the primary evidence supporting the notion that the 'expansion of the universe is accelerating'—certain stellar objects in other galaxies presenting a different apparent luminance than their redshift-based

distance estimate supports—can be far more economically interpreted by realizing that there are other causes of redshift than recessional motion. In other words, these stellar objects are much closer and much younger than otherwise believed. Second, newborn stars are concentrated within high redshift extragalactic objects.[128] The big bang interpretation of redshift data suggests thus that stellar evolution experienced a 'growth spurt' just after the big bang—the newborn stars being 'far away and receding fast'. A steady-state interpretation of the same redshift data instead suggests that stellar creation may be temporally uniform—occurring in 'quasars' that are not far and flying away, but within close, young, parent galaxies. Third, I would suggest that the microwave and radio emissions 'proving' the Big Bang theory could just as well emerge from a model in which galactic centers act as continuous matter recyclers, with old matter coming in and young proto-matter coming out, following an evolutionary path not dissimilar from that conjectured to have followed a singular 'Big Bang'. The aggregate radiation fields emerging from all galaxies everywhere and everywhen may be indistinguishable from that which we otherwise ascribe to a central explosion of energy 15+ billion years ago. Fourth, I will—I believe uniquely—speculate that the quantization in extragalactic redshifts observed by Arp and others may indicate that stellar evolution occurs in phases which may temporally correlate with matter aggregation, possibly correlating further among galaxies by virtue of equilibration effects over cosmic

5. time of the physical resonances among galaxies that may accompany such powerful events as galactic births. Sort of like the synchrony we see among the reproductive habits of coral reefs in Earth's oceans: near simultaneous release of spawn across vast distances. Then natural cycles of evolution follow, which remain roughly synchronized in time. In this view, light departing young galaxies might be attenuated in phased relation to a 'placenta' of 'amniotic' charged particles as yet unformed into new stars. If so, such a process might account for the observed quantization in extragalactic redshifts of quasars.

6. Reports emerging from recent astronomical observations indicate that the empirical measurements of 'black holes' do not match the

predictions of general relativity. Observations from four spacecraft have identified the inner edge of a spinning disk of material around a black hole about 5,000 light-years from Earth. The surprising results show that the disk is much farther from the black hole than astronomers expected.129 Separately, a recently measured X-ray/radio flare-up of the supermassive black hole at the center of one galaxy has raised similar questions. The measurements were, for the first time, able to pinpoint the location of this black hole to high precision. Investigators have reported that nearby orbiting stars indicate that the event horizon is 30,000 times larger than it is supposed to be for the measured black hole mass as predicted by current theory. And the nearby orbiting matter is 1,500 times farther away from the event horizon than it should be as predicted by current theory.130

7.  The widely reported reanalysis of the Large Electron Positron (LEP) Collider data at CERN is casting serious doubt upon the existence of the Higgs particle. According to a Dec. 5 report in the New Scientist: "The legendary particle that physicists thought explained why matter has mass probably does not exist. So say researchers who have spent a year analysing data from the LEP accelerator at the CERN nuclear physics lab near Geneva." This news has direct relevance to the work of a network of scientists who have uncovered a promising connection between the electromagnetic quantum medium (also called the zero-point field) and the origin of mass. Their approach—which we will review shortly—is significant because it does not require a Higgs field to explain the origin of mass.

8.  Fundamental questions have been raised about the prevailing model of stellar evolution by several decades of analysis of emissions from our own Sun. Dr. Oliver Manuel's research spanning more than three decades has gained stark confirmation from NASA's recent SOHO and TRACE missions, revealing that the Sun is very possibly made largely of iron, with a solid surface under the photosphere that rotates uniformly at all latitudes, from which energy and light result not mainly from hydrogen fusion, but from a superconducting iron or neutron core. Further, he and others are now suggesting, based upon

strong chemical evidence found in planets and asteroids, that our solar system was formed directly from a prior supernova, whose core became the core of our Sun. Recent data are accumulating rapidly in support of this startling new picture.[131]

9. There are several other reasons that ought motivate serious people to reconsider their assumptions concerning the solidity of the prevailing theoretical framework of physics[132], including unacknowledged and serious controversies concerning the empirical data allegedly proving Einstein's interpretation of the relativity of physics[133].

But perhaps as motivating as any single discrepancy between theory and measurement, a different interpretation of the empirical data that led to the development of Einstein's relativity is rapidly gaining ground—describing a paradigm of physics that possesses four vital qualities: it formally predicts the results of most key experiments that have validated Einstein's relativity, it suggests new classes of experiments that might distinguish between interpretations of existing data, it predicts none of the philosophically-challenged exotica described in the previous section, yet it suggests that remarkable technologies are possible whose relevance to humanity is difficult to overstate.

## The Struggle to Reexamine Einstein's Interpretation of Relativity

Scattered among clippings on the editing room floor of 20th century physics are several extremely interesting alternative theories explaining the empirical observations that led Einstein to his interpretation of the relativity of physics. References to many of these have been noted above, and they all today cry out for new hearings among physicists committed to science above science fiction. Each of them suggests that at least one major mistake of theory or interpretation is rooted somewhere in the apparatus of modern physics; the earlier the era of the hypothesized mistake(s), the greater the implications for science.

One of these alternative theories is set forth in Thomas Phipps insightful and well-articulated Heretical Verities, built around his study of Hertz's alterations of Maxwell's equations. Following detailed arguments against the plausibility of special relativity, Phipps steps up in support of Einstein's conclusion that no kind of 'ether' exists, and then expands upon Hertz's examination of the relativity of electrodynamics. He writes, "[Hertz] had a sound instinct that warned him not to include an unmeasurable quantity [ether] in his equations, so he borrowed an old hypothesis . . . to the effect that ether is dragged along by ponderable matter of any description . . . he would soon (I speculate) have switched to the opposite view, that ether can be entirely dispensed with . . . In that case he would have been in a position of strength to lock horns with Einstein on the rutting grounds of a 'true relativity theory'."[134] Leveraging Hertz's alterations of Maxwell's equations, Phipps concludes that the Lorentz contraction is fiction, and thus that special relativity's explanation of it simply compounds one grave fiction with another. He further bases this assertion on other grounds, two of the most important of which are: (1) the singular dimension of the alleged contraction—in the direction of motion—makes the math 'fragile' when mapping Nature, since no motion of any elemental body is truly linear, and (2) the very troubling fact that Lorentzian length contraction has never been experimentally verified except by circumstantial implication of the Michelson-Morley-type data. No one has ever empirically measured the 'shrinkage' of a ruler, because doing so proves to be no small challenge.

I find Phipps' arguments interesting and worth serious empirical examination. However, one nagging disagreement preoccupies my reading of his work: I do not believe that a physical contraction in the linear direction of the momentum of a body is necessarily theoretically 'fragile', when a non-kinematic (dynamic) interpretation of 'inertial motion' is considered. The meaning of this criticism will become apparent later in this monograph.

Another of the more compelling alternative theories of the relativity of physics emerged early in the 20th century, from Hendrick Lorentz himself—a scientist whose professional competence and contributions are praised by almost all modern theoreticians. As described above, Lorentz had a significantly different interpretation than did Einstein of the null result obtained by Michelson and Morley in their attempts to find a directional variance of the measured velocity of light. Lorentz believed that the speed of light was measured to be constant relative to the observer because the systems of charge (atoms) used to measure it—rulers and clocks—are also electromagnetic in nature, and thus physically contract and dilate in precisely the manner that yields the subjective measurement of a constant relative velocity c of arriving and departing electromagnetic impulses (light). In this view, the dimensions of space and time do not 'warp'. Rather, the shapes and internal motions of physical bodies vary with motion through, and properties of, a physical medium—yielding the apparent relativity of light speed and the invariance of the laws of physics we observe. But as compelling and simple as Lorentz's interpretation was, it could not compete with the almost magical allure of Einstein's geometrization of reality—particularly given the possibly excessive fascination with non-Euclidean dimensional mathematics carried in with the otherwise more balanced positivist winds then blowing through the atmosphere of physics.

As the years went by, a few restless theoreticians and experimentalists kept irritating peers by resurfacing questions about Einstein's relativity. However, in the absence of a simply performable, broadly replicable experiment capable of laying bare a glaring anomaly between Nature and the predictions of special or general relativity, the philosophical economy of Lorentz's physics failed to overcome the seduction of Einstein's math.

The acceptance of Einstein's interpretation of the relativity of physics soon became so widespread that empirical studies of inertial motion and gravitation were rarely attempted. After all, why would an aspiring physicist choose to waste valuable career time retracing an already-cleared trail of science? Expressing concern over this situation four decades later, noted physicist R. H. Dicke observed, "One distressing thing about the structure of General Relativity, from the first inception of the theory up to the present day, is the lack of contact of the theory with observational and experimental facts. This may be the reason that the theory is so strongly based upon philosophical arguments. The rather surprising thing is that a theory having this type of origin should become so firmly established. Certainly physicists must all agree that the primary basis for a theory should be observations, not philosophical arguments. As time has passed, General Relativity has become even more remote from observations. A large number of theoretical papers are published every year on General Relativity and most of them are motivated by formal questions, having little to say about any experimental or observational data."[135] "It is strange that the gravitational-inertial field which is fundamental to all physics, particularly in its inertial aspects, is but poorly known from an experimental or observational view point. There has been very little significant experimental work done on gravitation in the past fifty years."[136]

It was, in fact, Dicke who picked up where Lorentz left off. He published a landmark paper in 1961 exploring what might have happened if Einstein had appeared on the stage of science just a few years later than he did. In an imagined scenario he titled, It could have happened,[137] Dicke wrote, "Let us imagine that the year is 1906, that Lorentz's electron theory has been exhibiting its phenomenal series of successes, that Einstein is 15 years younger than he actually was, and that two remarkable observations are made. First it is discovered that Lorentz's formula for the velocity dependence of the electron's mass is correct to considerable accuracy. Second, it is observed that a star image in the vicinity of the eclipsed sun is shifted from its normal position . . .

"Lorentz had skillfully woven an intricate theoretical fabric in which seemingly all parts of physics were related to each other. Optics, heat, the dielectric properties of matter, electrical conduction, all had been related to the electron and its motion. The null result of the Michelson-Morley experiment was related to the length contraction which resulted from the motion of electrically charged matter through the ether. Lorentz, following earlier suggestions of J. J. Thomson and M. Abraham, was even able to account for the inertial mass of the electron as an electromagnetic effect . . .

"Lorentz quickly enters the lecture hall and announces that he has received word from Germany of two important observational results. First, Bucherer has found an error in the experimental results of Kaufmann, and Lorentz's formula for the mass variation of the electron is substantiated with considerable precision . . . Second, photographs of the 1905 eclipse of the sun show that light is deflected by the sun and that the deflection is twice what one would expect naively from Newton's corpuscular picture of light. (It is hardly necessary to remark that historically the observation of the deflection of light did not occur until 1919 and then with no great accuracy.)

"Lorentz had first obtained this news in a telegram the preceding evening and had spent the whole night thinking about the implications of the observation of the deflection of light. It had been immediately clear to him that these observations required the assumption of a concentration of the ether, about gravitating bodies. He had earlier given some thought to the possibility of such a condensation in connection with the problem of stellar aberration and had considered it unnecessary . . .

"Now he had considered more carefully the implications of a greater ether density in the vicinity of a gravitating body. He noted that a charged particle would be attracted toward regions of the ether with a higher dielectric constant. Finally it becomes completely clear, gravitation, the one physical phenomenon which he had been unable to incorporate into his theory, was also electromagnetic. A body falls toward the earth because the charged particles, of which the body is composed, tend to move into a region where the dielectric constant of the ether is greater.

"Lorentz spent the remainder of the night constructing the obvious theoretical formalism which would incorporate these ideas. He found that there was an almost unique theoretical path through the formalistic jungle and by morning he had a theory, complete in its principal features."

Throughout this alternate history of early 20th century physics, Dicke weaves a detailed presentation of the formal equations which, he conjectured, Lorentz would easily have derived. He continues, "The fact that Lorentz's theory was able to account for the enigmatic rotation of the perihelion of Mercury's orbit, in addition to accounting for the deflection of light by the sun, made it immediately acceptable. When some years later it was found that the gravitational red shift predicted by the theory existed quantitatively correct, the theory was more firmly established than ever . . .

"It is now fifteen years later. The local Lorentz invariance of the above equations is recognized as a useful tool and is widely used, but Lorentz's interpretation of the invariance property, as indicating the universal electromagnetic character of matter, is the one universally adopted. The Fitzgerald contraction of moving rods and the time dilatation of moving clocks are interpreted as electromagnetic effects resulting from their motion through a fixed ether. The fact that an apparatus cannot be used to measure locally the velocity relative to the ether is regarded as a compelling argument for the ultimate electromagnetic character of matter.

"It is obvious that Lorentz would have recognized that a geometry based on a local measure of length and time, with its locally distorted meter sticks and clocks would be non-Euclidean. Thus, Lorentz would have ultimately recognized that his equations of motion of matter could be expressed as geodesic [path of motion] equations, in a Riemannian metric ['warped' space-time dimensions]. However, he would not have been likely to favor this mode of expression, which he would have considered unphysical.

"It is doubtful that the young physicist, Albert Einstein, as ingenious as he was, would have concerned himself with gravitation under these conditions. However, if he had produced his generally covariant purely geometrical theory based on the equivalence principle, it seems likely that it would have been considered unphysical and a complicated way of getting approximations to old results. While his use of a Riemannian geometry to describe gravitation as a purely geometrical effect would have been regarded as an interesting though complicated trick, the Einstein field equations would have been considered wrong for they
would be found to give the 'wrong' answer. By 'wrong' here one means equations which differ somewhat from those of Lorentz's theory, the already established, hence 'correct', theory."

Dicke's moral to his story: "The seniority of a theory is not a proper basis for its acceptance. All theories should be considered and the acceptance of any theory should be tentative only. That theory should be favored which in the simplest way accounts for the most experimental facts within the framework of the most satisfactory philosophy. If in balancing these requirements, a conflict should develop between the demands of observations and the demands of philosophy, it should be resolved in favor of the observations . . .

"[My] theory . . . seems to account in a satisfactory way for the observations. Especially satisfactory is the fact that the 2nd order effect, the planetary perihelion precession, agrees with the observations . . . It is also an esthetically satisfactory theory, if one is willing to accept the notion of an absolute space. For example, it can be noted that by treating gravitation as an electromagnetic phenomenon, essentially all parts of physics are reduced to some aspect of electromagnetism." As with Einstein's general relativity, Dicke's theory did not describe the specific mechanism by which a gravitational field is established, but rather the general physical nature and mathematics of its action upon matter.

Yet another example of the opportunity to reinterpret the meaning of relativity has been introduced by [Jefimenko]. In his 1997 book Retardation and Relativity, he concisely reveals how the key problem Einstein attempted to address in special relativity—namely, the preservation of the laws of electrodynamics regardless of relative motion among bodies—is resolved far more simply by properly accounting for the effects of the speed-of-light delay of all electromagnetic influences among bodies. He argues quite convincingly that the physical effects described by Lorentz and Einstein—apparent length contraction, apparent time dilation and apparent mass increase of moving objects—do not really exist in those objects. Rather, they exist only within the measures of electromagnetic images emitted by those moving objects. The effects of an object's motion on its pattern of electromagnetic emissions, he claims, can account for every concrete observation otherwise confirming Einstein's special relativity. All one must accept in exchange is the existence of some kind of absolute medium.

Whatever cliffs must be scaled to reach the summit of physical truth, this climber finds these latter footholds more rugged than the prevailing position of physics today. And from this vantage point, higher footholds are now apparent and within reach.

## An Emerging View of Mass in Motion

From studies of the quantum mechanical behavior of fundamental particles, it has long been known that 'space' appears to continuously jostle charged particles, even at zero temperature. This phenomenon—known as zitterbewegung—is fundamentally related to the nature of the quantization of electromagnetic interactions. In 1967, the great Russian physicist Andrei Sakharov conjectured that gravity may have its origin as a consequence of this energetic influence pervading space. He suggested that gravitational attraction might be an induced bias in the mutual interactions of charges experiencing zitterbewegung.[138]

Physicist Harold Puthoff writes, "Although speculative when first introduced by Sakharov in 1967, this hypothesis has led to a rich and ongoing literature on quantum-fluctuation-induced gravity that continues to be of interest. In this approach the presence of matter in the vacuum is taken to constitute a kind of set of boundaries as in a generalized Casimir effect, and the question of how quantum fluctuations of the vacuum can reproduce under these circumstances the action and metric of Einstein relativity has been addressed from several viewpoints."[139]

Compatible with Dicke's interpretation of relativity, and building upon the work of Sakharov and others following this thread in 1989, Puthoff presented a formal hypothesis for the origin of gravitation as such an effect: "Sakharov has proposed a suggestive model in which gravity is not a separately existing fundamental force, but rather an induced effect associated with the zero-point fluctuations (ZPFs) of the vacuum, in much the same manner as the van der Waals and Casimir forces. In the spirit of this proposal we develop a point-particle-ZPF interaction model that accords with and fulfills this hypothesis. In the model gravitational mass and its associated gravitational effects are shown to derive in a fully self-consistent way from electromagnetic-ZPF-induced particle motion (Zitterbewegung). Because of its electromagnetic-ZPF underpinning, gravitational theory in this form constitutes an 'already unified' theory."[140]

Although problems remain in his 1989 formulation, Puthoff's general approach to the question of gravity's mechanism further opened the door that closed upon electromagnetic theories of gravitation and inertia nearly a century ago.

Continuing along related lines of reasoning, in 1994, using a semiclassical technique in physics known as Stochastic Electrodynamics (SED), Lockheed astrophysicist Bernhard Haisch, mathematician Alfonso Rueda and Puthoff published the hypothesis that inertia may also originate in interactions between the electromagnetic zero-point field of the quantum vacuum and the charges constituting matter.[141] Their analysis yielded a startling insight: that Newton's equation of motion, F=ma, heretofore regarded as a postulate of physics, might be derivable from Maxwell's equations as applied to the electromagnetic zero-point field.

Haisch writes, "This led to a NASA-funded study beginning in 1996 at the Lockheed Martin Advanced Technology Center in Palo Alto and the California State University in Long Beach. That study found the more general result that the relativistic equation of motion could be derived from consideration of the Poynting vector of the zero-point field in accelerated reference frames, again within the context (and limitations) of SED.

"It is well known that an accelerating observer will experience a bath of radiation resulting from the quantum vacuum which mimics that of a heat bath, the so-called Davies-Unruh effect . . . For an accelerated object moving through the vacuum the zero-point field will yield a non-zero Poynting [electromagnetic momentum] vector. Scattering of this radiation by the [charges] constituting matter would result in an acceleration-dependent reaction force that would appear to be the origin of inertia of matter.[142] In the subrelativistic case this inertia reaction force is exactly Newtonian and in the relativistic case it exactly reproduces the well known relativistic extension of Newton's Law. Both the ordinary, F=ma, and the relativistic form of Newton's equation of motion may be derived from Maxwell's equations as applied to the electromagnetic zero-point field."[143]

In papers published in 2001, Rueda and Haisch have formally developed this theory to account for the force of weight felt by objects in a gravitational field.144 "GR declares that gravity can be interpreted as spacetime curvature. Wheeler coined the term geometrodynamics to describe this: the dynamics of objects subject to gravity is determined by the geometry of four-dimensional spacetime. What geometrodynamics actually specifies is the family of geodesics—the shortest four-dimensional distances between two points in spacetime—in the presence of a gravitating body. Freely-falling objects and light rays follow geodesics. However when an object is prevented from following a geodetic trajectory, a force is experienced: the well-known force called weight. Where does this force come from? Or put another way, how does a gravitational field exert a force on a non freely-falling, fixed, object, such as an observer standing on a scale on the Earth's surface? This proves to be the identical process as described in the quantum vacuum inertia hypothesis.

"In the SED approximation, the electromagnetic quantum vacuum is represented as propagating electromagnetic waves. These should follow geodesics. It can be shown that propagation along curved geodesics creates the identical electromagnetic momentum flux with respect to a stationary fixed object as is the case for an accelerating object. This is perfectly consistent with Einstein's fundamental assumption of the equivalence of gravitation and acceleration. An object fixed above a gravitating body will perceive the electromagnetic quantum vacuum to be accelerating past it, which is of course the same as the perception of the object when it is doing the accelerating through the quantum vacuum.

Another useful intuitive picture is to imagine the downward deviation of tangential light rays near a gravitating body resulting in a net downward force, somewhat analogous to radiation pressure, on a fixed object.

"Thus in the case of gravity, it would be the electromagnetic momentum flux acting upon a fixed object that creates the force known as weight . . . Since the same electromagnetic momentum flux would be seen by either a fixed object in a gravitational field or an accelerating object in free space, the force that is felt would be the same, hence the parameters we traditionally call inertial and gravitational mass must be the same. This would explain the physical origin of the weak principle of equivalence."[145]

The recent work of Haisch and Rueda concerning the nature of weight is compatible with Puthoff's recent reintroduction and extension of Dicke's interpretation of general relativity. Based also on the prior work of world-renowned physicist Huseyin Yilmaz, in a landmark 2001 paper, Polarizable-vacuum approach to general relativity, Puthoff thoroughly reviews how the 'space-time curvature' of general relativity is more economically interpreted as the variation of the properties of a physical medium, much like the variation of an index of refraction. He writes, "Topics in general relativity (GR) are routinely treated in terms of tensor formulations in curved spacetime. [The new approach is based] on treating the vacuum as a polarizable medium. Beyond simply reproducing the standard weak-field predictions of GR, the polarizable vacuum (PV) approach provides additional insight into what is meant by a curved metric."[146]

Combining Puthoff's and Haisch's hypotheses, we can assert with no less empirical confidence than today supports Einstein's interpretation of relativity that gravitation represents the refraction of 'inertial' momentum. When a body departs from motion along a geodesic, the 'weight' (if the body is resisting gravity) or 'inertia' (if the body is resisting applied impulse) it feels can be understood, literally, as a flux of momentum—an impulse—traversing the body, resulting from the interaction of the body's charges with the medium through which they are moving. This thoroughly conforms to our common sense experience.

It is useful to summarize the implications of these different interpretations. We will refer to the competing theories as GR (canonical general relativity) and PM (polarizable medium interpretation of general relativity) respectively.

Concerning the Equivalence Principle:

> In canonical GR theory, 'general relativity' means that mechanics and electrodynamics are locally measured to be invariant because the dimensions of space and time, relative to which all such phenomena are locally measured, are warped by matter. Hence, gravity is a space-time warp. Hence, the experience of gravitational force is equivalent to the experience of departure by mass from straight-line motion in its space-time dimensionality—departure from pure inertial motion along a geodesic.

> In PM theory, 'general relativity' means that mechanics and electrodynamics are locally measured to be invariant because they scale with and are governed by the local properties of a physical dielectric medium, which properties change in the vicinity of matter. Hence, gravity is an asymmetry (a so-called "K=/=1" state) in the energetic medium that causes ruler-clock warps with respect to Euclidean space and proper time (the so-called "K=1" state). Hence, the experience of gravitational force is equivalent to an asymmetric experience of the energetic medium—departure from pure inertial motion.

Concerning the Michelson-Morley experiment:

> From the point of view of special relativity incorporated into GR theory, the Michelson-Morley experiment established the absolute constancy of the relative speed of light regardless of observer motion.

> From the point of view of PM theory, the Michelson-Morley experiment established that the ruler-clocks used to measure the speed of light experience the same dielectric medium effects as the light they measure, yielding the appearance of constancy in the speed of light.

Concerning the nature of the gravitational force:

> In GR theory, a motion along a geodesic means no force is operating to influence inertial motion. Force is thus explained ontologically as

a dimensional warp, which warp is assumed to be the ontological fundamental. It is from this swapping of epistemological and ontological constructs that physicists have conjured black holes, wormholes, big bangs, time travel, and no absolute concept of space or time.

➢ In PM theory, all inertial motion follows the geometry of balance in continuous force from a physical medium. Force is thus explained ontologically as fundamental, or related to a like fundamental, such as energy. Hence PM theory predicts no black holes, no wormholes, no time travel, no big bang, but useful concepts of absolute space and time (Length and Time in a K=1 region).

Puthoff has referred to PM as GR "au naturale" meaning, simply, that PM is, in a real sense, at least as generally relativistic as canonical GR. PM says that the speed of light through the medium changes, along with the material objects that measure it, such that relative c is always measured locally, and that physical objects are in fact changing shape and physical clocks are dilating, within an energetic medium, whose objective dimensionality is Euclidean. All of mechanics and electrodynamics are thus physically, not simply mathematically, generally relativistic.

As Dicke and Puthoff have progressively shown, every historical experimental result supporting Einstein's equations of relativity remain formally predicted, yet Einstein's interpretation of gravity—warped dimensions of space and time—is replaced with a far more logical, efficient and technologically-relevant interpretation: gravity arises by virtue of the variability of the properties of a physical medium extant in Euclidean three-dimensional space and uniformly flowing time. Yet PM theory also predicts that in extreme limits—on very small scales and in very intense gravitational fields—an experimentally-detectable divergence emerges between the predictions of the two formalisms, pointing the way towards fertile ground for both laboratory and astrophysical study.

Hanging in the balance of this debate are such epic questions as the nature of the dimensions of space and time, the nature of the medium, and the nature of mass and force. But the relevance of the debate extends beyond understanding alone, for it opens the door of possibilities to apply deeper insights into gravitation and inertia for new kinds of urgently needed clean energy and propulsion technologies. Haisch writes, "What this view adds to physics is insight into a specific physical process creating identical inertial and gravitational forces. What this view hints at in terms of advanced propulsion technology is the possibility that by locally modifying either the electromagnetic quantum vacuum and/or its interaction with matter, inertial and gravitational forces could be modified or even nullified."[147]

## *Glimpsing the Mechanism of Gravitation and Inertia?*

Scattered among physics laboratories around the world are courageous teams exploring these questions in experimental terms. One of these experimentalists is NASA scientist Creon Levit, who writes, "Today it is most often taught that electrodynamics is a complete theory—that there is nothing new to be found, and certainly nothing amiss, at the foundations of electricity and magnetism. However, there actually are many areas of electrodynamics that still contain mystery, paradox, and surprise. The troubles in classical electrodynamics are old and are many. Some examples are: the predicted electric field around an electron becomes infinite as one approaches the electron more closely, static (unchanging) electromagnetic fields can contain arbitrarily large quantities of momentum, and electrons should begin to accelerate under the influence of electromagnetic fields before the fields even reach them. Maxwell's equations, the cornerstone of mainstream theoretical and industrial electrodynamics, predict electromagnetic waves that travel backwards in time. (For a summary, see Rohrlich).

"Quantum electrodynamics (QED), which was supposed to resolve these difficulties, does so at best only partially, and introduces worse problems of its own. R. P. Feynman, who received the Nobel prize for his formulation of QED, described the theory as a 'shell game.' Problematic consequences of QED arise from its postulated infinite unobservable virtual particles (and hence infinite self energy, also not observed) in every region of space, however small. Additional infinities need to be subtracted from the existing infinities to produce results in accord with experiment. QED also carries with it the vast interpretational difficulties of quantum mechanics itself.

"Additional insight and possible resolution of these problems can be obtained via a number of alternate formulations of classical and quantum electrodynamics. In addition to stochastic electrodynamics [SED, mentioned above] we are particularly interested in the 'neoclassical' electrodynamics of Barut, Jaynes, Lamb, Crisp, and Marshall. These self-consistent, quantitatively accurate, and intuitively clear formulations of electrodynamics appear to predict most if not all of the results of QED, without the need for virtual particles, infinities, and quantum paradoxes.

"The notion of extended (non-pointlike) electrons and other elementary particles is also of significant interest to us. Using these models (e.g. Barut, Bostick, Reccami, Hestenes, Bergstrom, etc.) we can predict or at least explain many of the properties of elementary particles that are simply postulated ex nihilo in QED (for example: spin, Zitterbewegung, uncertainly, and particle stability, or lack of it). Closely related to this are hydrodynamic analogues for elementary particles (Sudarhsan, Nambu, Fadeev, etc.). In these models, particles are described as coherent structures (e.g. vortices), excitations, or topological defects in an underlying three-dimensional superfluid vacuum. These hydrodynamic models of particles and electromagnetism may also explain the other forces of nature (strong, weak, gravitation, etc.) as higher order effects via the same underlying mechanisms. They constitute, or at least suggest, alternatives to the standard model, and may even be mathematically isomorphic to the objects in superstring theory.

"Much of the progress in physics has come to pass via questioning and extending conservation laws. Clearly, in order to obtain efficient propellantless propulsion, or 'free' energy production, the principles of conservation of momentum and energy, respectively, may have to be extended or amended with a covering theory, much as special relativity's $E=mc^2$ extended and amended the prior 'law' of conservation of mass. Interestingly, as Gray points out, conservation of electric charge has been tested with far less accuracy than have the more exotic conserved quantities of elementary particle physics. Nonconservation of any of these quantities, or even improved formulations of their definitions (e.g. electromagnetic momentum), could lead to substantial progress in both theoretical and applied electrodynamics.

"There are nagging questions in electrodynamics, many of them deceptively simple, some left over from the earliest days, which are not yet resolved. For example, the relationship between magnetism and rotation (see Sirag, Wesson, Gray), homopolar induction, the debate over moving field lines, and the correct form of the expression for the force between current elements. Basic stuff. Still unknown.

"There are many formulations of electrodynamics that are not necessarily equivalent to the Maxwellian electrodynamics that dominates today. Ampere's formulation was actually preferred by Maxwell to his own, since it described a larger class of phenomena. The electrodynamic theories of Ampere, Neuman, Weber, Grassman, as well as other alternatives by more modern physicists, make predictions where Maxwell theory is silent. (For surveys see O'Rahilly, Graneau, Assis). Some electrodynamic phenomena observed in the laboratory and elsewhere are explainable using these alternatives, but apparently not using Maxwell (see Barret and Grimes for examples).

"We are actively researching, on the blackboard and in the laboratory, electromagnetic configurations and applications suggested by these alternative theories and extended definitions. We have some promising leads and have made some significant progress, but there is a great deal of work that remains to be done."[148]

# Chapter 6
# A Tale of Two Ontologies

*"And when [a discipline of science] can no*
*longer evade anomalies that subvert*
*the existing tradition of scientific*
*practice—then begin the extraordinary*
*investigations that lead the profession at*
*last to a new set of commitments, a new*
*basis for the practice of science. The*
*extraordinary episodes in which that*
*shift of professional commitments*
*occurs are the ones known . . . as*
*scientific revolutions."*[149]

*—Thomas Kuhn*

An epic discussion is rising today within the halls of universities, research laboratories, dozens of major science journals, and among scores of communities across the Internet. It involves the deepest questions of science. It reflects an increasingly-held view among scientists that physics unknowingly hit a roadblock somewhere in the past century, the consequences of which may stagger our imaginations.

Consider the possibility: a few pivotal, understandable, ever-unavoidable approximations are scattered among the theories of 18th-20th century physics. Consider that they are sufficiently subtle that they have avoided pronounced experimental detection through decades of modern research. Consider that the implications of these deficiencies in self-comprehension may represent the primary obstacle to the emergence of knowledge that could send us to the stars, permanently cleanse our air and water, bring well-being to humanity as a whole, and ontologically reconnect us with the Nature from which we spring. Consider that these defects may be perpetuating a collective myopia among 6+ billion human beings and all of the institutions we've created—a myopia of epistemological and ontological import.

In the spirit of promoting open-minded conversation between the 20th century physics we know and the 21st century physics we don't, allow me to attempt to generalize principles partially suggested by the above-mentioned alternative, emergent theories of physics. Allow me to share a tale of two ontologies, in terms comprehensible to any philosophically-competent college graduate equipped with a basic understanding of physical science. The first framework yields the theoretical paradigm sustained by mainstream physics today. The second framework describes one possible approximate generalization of a different vision of Nature.

## Space-Time: the Dimensions of Forms in Motion

In the canonical view of mainstream physics today, the fundamental concepts of space, time, force, and mass may be summarized as follows:

**Space**: a dimensionality of reality measurable by rulers, whose non-Euclidean geometry is the basis of non-linear motion.

**Time**: a dimensionality of reality measurable by clocks, and whose non-uniformity is the basis by which the laws of physics are strictly relative.

**Force**: a label historically applied to the effect of non-uniform space-time upon the motion of mass.

**Mass**: the as-yet-unexplained reaction force experienced by a body when it deviates from linear motion in the geometry of space-time, whose origin warps the dimensions of space and time.

With these mental pictures in mind, consider an alternative paradigmatic vision of these fundamental concepts of physics.

## Potentum: the Medium Comprising Fundamental Forms in Motion Space

As suggested previously, the work of Ampere, Maxwell, Lorentz, Dicke, Sakharov, Yilmaz, Puthoff, Haisch, Rueda and others, along with the work of those pursuing 'electrodynamic superfluid' models of reality argue that physics revisit the question of the nature of 'space' with an expanded theoretical and experimental toolbox.

In my review of pre-scientific and classical concepts of space, I observed that notions of space have progressively differentiated into two general and competing ideas: (a) space is a dimensionality in which all things exist, and (b) space is a dimensionality of which all things are made. In the former case, space is usually viewed as an idealized, unbounded, three-dimensional void in which matter exists, moves and interacts. In the latter case, space is usually viewed as a kind of 'metric' or 'field', whose dimensions are not "flat" in the common sense we experience, but rather 'warped' to cause and reflect the ways matter moves and interacts. In this latter view, it is now often speculated that matter particles themselves are knots or strings in the metric or field of space (or space-time).

I will hypothesize that we can evolve usefully our understanding of space if we formally distinguish the aspects of the notion which give reason for natural philosophers to choose either concept (a) or (b) in defining space. In doing so, it becomes apparent how both concepts of space play essential roles in physics. Consider a clarification of the nature of space through the following four-step reasoning process.

First, we will postulate that all measurements are in themselves subjective, relative constructs of observers, and that subjective, relative conceptions are not ontologically equivalent to the nature of the physicalities measured. This opens a way to distinguish the objective and subjective constructs of spatiality. The objective concept of 'space' can be seen as a construct shared by all observers by which we compare subjective measures of all physical systems. The concepts of observed 'volume' or 'velocity' can be seen as specific subjective measures of space occupied or traversed by a physical system measured in local, relative terms.

Next, let us consider that the variability of shape, volume, velocity and path of a physical system is a basis by which the relativity of physics operates, and that this variability is not the result of a 'warp' in Euclidean dimensionality. Rather, we may consider that relativity operates by virtue of the local variability of a physical, energetic medium relative to idealized, Euclidean space, and/or of the subjective experience of this medium by a body in motion through it. In his polarizable medium interpretation of relativity, Puthoff defines idealized measures of spatiality and temporality as the objective standards enabling 'absolutization' of local determinations of shape, volume, velocity and time. In his theory, the idealization of space and time is a ruler/clock measure in a region of space where K=1.

A simple analogy makes this conceptual structure clear: consider a molecule M of water in the ocean. M is, naturally, subject to a certain surrounding pressure P from the molecules surrounding it. It is intuitively and empirically clear that P increases with the depth of M. The deeper in the ocean M resides, the more surrounding pressure it experiences, slightly shrinking its spatial dimensions and changing its temporal rate of activity. Yet from M's perspective, there is no net 'force' moving it this way or that, and the internal functions of M continue regardless of its depth. Indeed, because of the relative invariance of its physics, at least within certain limits, M does not 'know' what depth it may occupy in the ocean, or even that it is subject to a pressure P.

Consider the possibility that the physical mechanism—otherwise called 'space'—by which physics' measures of spatiality, motion and the temporal rate of action scale with perfect subjective relativity, yet remain comparable to an ideal objective standard, is in fact akin to a 'universal pressure'—an omnidirectional impulse. Such a 'quantum ether' would appear undetectable directly because we are made of it.

From this sequence of logic, physics' notion of space differentiates into two distinct concepts.

> We measure spatiality locally with rulers knowing that our local rulers may be physically modified if compared to the idealized ruler—a ruler at rest with respect to the Cosmic background, far from gravitating matter.
> Everywhere there exists a kind of 'pressure', or continuing force, which sustains and acts upon all bodies always. The bodies formed within this 'pressure' can be measured relative to each other, but also relative to the empirical spatiality established by all objects in the observable Universe—a good placeholder for which is likely the topography of the Cosmic background.

Might it be that rejected 19th century conjectures about an 'aether' were misguided because they assumed some kind of corpuscular constituency defining the medium? Might it be that the problems attendant to such a view—how the inertia of mass could appear undisturbed yet actively sustained by a corpuscular 'space'—were interpretive fictions?

Might it be that the space medium is, in fact, a continuum of omnidirectional pressure, purely and simply 'push', whose only topographic variations are density gradients which we call radiation, and in more complex self-sustaining forms, matter? What simpler ontology offers broader explanative power than an infinite medium of pressure differentiating only in traveling and self-interacting patterns of density?

Might it be that we humans are like fish in a Cosmic ocean whose electrodynamic pressure—the 'water' we call 'space'—we cannot sense, because we are made of it? Might we have mistakenly characterized local variations in this

pressure as 'dimensional warps' in our attempts to geometrize the pressure asymmetry—gravity—and the geometry of
interaction with it—inertia—that keep us floating near our Cosmic shore?

## Time

As reviewed above, modern physics' definition of time—that which clocks measure—is ambiguous at best. Yet the quantity t occupies a central role in virtually every equation of action in modern theory. So there is very good reason to consider an alternative concept of time. Indeed, as renowned physicist John Wheeler puts it, "Should we be prepared to see some day a new structure for the foundations of physics that does away with time? . . . Yes, because 'time' is in trouble." Let me offer what I believe is a unique view on the nature of time.

In examining the nature of all macroscopic physical clocks, two interesting facts become apparent: (1) without exception, clocks enumerate the recurrence of a spatial geometrical configuration of a system in motion, and (2) without exception, the unit of denomination of each enumeration is defined in terms of a microscopic reference clock. For example, a pendulum clock enumerates the recurrence of the pendulum arm crossing a certain position in its travel. A light clock enumerates the recurrence of electromagnetic impulses striking a photodetector. In both cases, we quantify the duration between enumerations in units of 'seconds' or some such measure. But what is the meaning of a 'second'? Is it not a certain enumeration of the motion cycles of a reference clock—in that case, 9,192,631,770 oscillations of a cesium atom?

In simplest terms, I assert that 'time' is, in fact, the cyclicality in the motion of coexistent physical bodies—a concept not far from the
intuition voiced by Aristotle millennia ago. If so, then a 'duration' is nothing more than an enumeration of geometrical cycles in the motions of matter. No more, no less. Our concept of 'past' represents, literally, the collection of cycles we remember along the trail of our path of motion. Our concept of 'future' represents the collection of cycles we project to occur along our forward path of

motion. The 'present' represents all that is actually ever real: co-existing bodies continuing in co-evolution through co-motion. Thus, among presently-moving bodies (reality), we humans carry with us information (memory) about past motions (history) that enables us to anticipate future motions (prediction).

In this view, time as we know it simply does not exist. Rather, motion exists, and counts of its cycles are what we call 'durations'. Interestingly, the concept of an eternal Universe becomes much easier to grasp: perpetual co-motion of bodies.

Let's review this definition of time a bit more carefully. Consider the view that both our experiential sense and our measurements of 'time' represent local enumerations of the spatial coordination of geometrically ordered, spatially nonlinear cycles in the motion of physical systems. Since we cannot easily measure time by directly counting the oscillations of atoms, we express measured durations in everyday terms as multiples of a reference unit which itself is such an enumeration. Thus, durations measured by our minds and everyday clocks are, in fact, the product of two enumerations: the first factor of the product reflects the explicit counting of the macroscopic clock motions we are measuring, while the second factor reflects the implicit counting of the motions of the microscopic reference clock defining the denominating unit of duration.

In this view, the dimension of 'time' today incorporated into physics equations is not a 'line-like' dimension, nor is it an 'arrow of time'. Rather, time is a count-like dimension, whose enumerative coordination is spatially determined by a pattern in relative geometry of motion. Thus, in measuring 'time' in real physical systems moving within a real physical medium, the second factor of the product must be considered subject to the physical reality extant at the chosen frame(s) of reference throughout the measuring process. Thus, a local clock can yield accurate relative measures of clock cycles, but does not necessarily produce a measure consistent with measurement taken elsewhere in the Universe (e.g. cesium atoms will oscillate differently in relation to the local properties of the medium and their motion through it), and though such clocks may run at different rates, they are nonetheless obviously, simultaneously existent. A future concept of a "universal clock" may identify a more perfect, standard unit of duration whose underlying enumeration of cycles of motion is derived from an ideal and/or real fundamental physical system determinable within the frame of reference most common to all physical systems, such as a clock far from any gravitating body at rest with respect to the cosmic background. In context of the polarizable medium interpretation of general relativity, the reference clock is a clock in a region where K=1.

Of interest to the theoretician may be this notion: by employing knowledge of local motion and properties of the local medium in ratio to the reference conditions used to measure a 'universal clock', we can scale the product of these factors, thus finding an alternative way to mathematically express the physical effect characterized as 'time dilation' within the equations of Einstein's relativity. While Newton's notion of absolute time may not be fully restorable, we may yet become able to measure durations in terms arbitrarily close to such a universal frame of reference, limited only by our empirical knowledge of the Cosmos.

If this line of reasoning is valid, relativity's assertion that local clocks experience 'time dilation' depending upon motion and gravitation is replaced by the assertion that local clocks experience local clock dilation depending upon the local properties of the medium in which they exist and their motion through it. The dimension of 'time'—the enumeration of geometrical cycles—does not dilate. It is the spatiality and/or pattern of motion of the clock—whose geometrical cycles of motion we enumerate—that may dilate. All the 'twin paradoxes' and 'time travelers' descending from Einstein's relativity are revealed to have different interpretations: the dimension of time—enumeration of cycles—is not dilating; rather material clocks are cycling in different motion patterns, and while they may thereby "age" at different relative rates, they are simultaneously existent. Thus, our common sense of simultaneity in existence is reestablished.

Also of interest to the theoretician, it may be possible to express this different notion of time in terms of the geometry of the nonlinear path of motion of fundamental bodies. If one assumes that the groundbreaking theoretical interpretations of Dirac by Hestenes and Shaffer are valid, then an electron follows (or is a vortex with) something like a helical trajectory in motion through the medium. Though these scientists do not appear to assert as much, I would speculate that their interpretation has fundamental relevance for our understanding of the nature of time. In my view, cycles of electron motion—turns of the helix—can represent 'ticks' of a clock, whose path length (in the ideal case and—presumably—relative to the empirical Cosmic background) is given by,

$$t = n\sqrt{((2\pi r)^{\wedge 2} + \lambda^{\wedge 2})}$$

where r is the radius of the helix, $\lambda$ is the distance between cycles, and n is the number of cycles enumerated, yielding a value t with the dimension of meters. Consider the equation of velocity expressed in these terms, through which velocity becomes seen as the ratio between the linear measure of distance traversed and the nonlinear length of the path actually followed, yielding a value v with dimensions meters/meters.

Let's take it further: acceleration becomes seen as the change in the ratio between the linear distance traversed and the nonlinear length of the path actually followed, divided by the nonlinear length of the path actually followed, yielding a value a with dimensions meters/meters$^2$. This concept would seem to suggest many opportunities for potentially fruitful theoretical development, some of which we will explore shortly.

In any case, as can readily be seen, the primary innovations in the proposed conception of time are these: the dimension of time we have used in physics to date may be the dimension of enumeration of spatial cycles of nonlinear motion, and, if so, this distinct dimension of time might be eliminable from—or otherwise replaced with a more empirical measure of the absolute nonlinear path of motion in—equations of physics. The implications of this, if valid, fundamentally affect each of the other metaphysical concepts we've explored in this monograph. All equations of physics expressing relations among measurements are epistemologically and ontologically incomplete if they include an empirically imprecise notion of time.

One might ask, 'This is interesting, but how can a purely spatial concept of time adequately express our sense of its flow? If this new notion of time is true, from what does our temporal sense emerge?'

In my view, the key to comprehending the sensation of temporal existence is to distinguish the mental picture of 'temporal' from that of 'existence'. Consider that the sense of 'existence' is not rooted in the concept of 'time'—that a 'physical' thing is not 'physical' because we can count its cycles of motion with a mental ruler called 'time'. Instead, consider that a thing is 'physically existent' if it exhibits any quality of 'pressure', or 'potential', to act upon its kind.

Indeed, among all concepts of 'existence' in human thought, what simpler 'substance' than 'pressure' is capable of acting upon its kind? Might the existential sense know itself through refined concepts of force and mass?

## Force

As can be seen from our discussion of classical concepts of 'force', the more extreme faction in the positivist trajectory of physics has successively removed ontological significance to whatever reality underlies the term. Max Jammer, in the concluding words of his thorough review of the subject, writes, "These considerations have brought us to the brink of present-day research in theoretical physics. If it were possible to work out a unified field theory that subjects electromagnetic and possibly also nuclear forces to a similar treatment as gravitation, it would lead us to a final stage in the history of the concept of force. While the modern treatment of classical mechanics still admitted, tolerantly, so to say, the concept of force as a methodological intermediate, the theory of fields would have to banish it even from this humble position."[150]

Indeed, Mach's hope would seem all but fulfilled.

Yet from time to time, competent philosophers and physicists have asserted that our common sense concept of 'force' represents a deep ontological root, whose nature should not be confused with the necessary mathematizations of its relations. Professor Mary B. Hesse writes, " . . . the concept of force in contemporary physics plays the role of a methodological intermediate comparable to the so-called middle term in the traditional syllogism . . . Newton defines force as that which produces acceleration . . . Hence force can be said to be an absolute cause of the acceleration. Nothing in the theory of relativity has altered this situation, for we do in fact distinguish inertial from accelerated frames, however the difference between them may be explained . . . although the mathematical sign of force may be a mere intermediate and eliminable term[,] the interpretation of the term is not." True, Bigelow, Ellis, and Pargetter agree when they write, " . . . forces should not be eliminated—just differently construed. For the effect of elimination is to leave us without any adequate account of the causal relationships forces were postulated to explain. And this would remain the case, even if forces would be identified with some merely dispositional properties of physical systems. In our view, forces are species of the causal relation itself, and as such, have a different ontological status from the sorts of entities normally considered to be related as causes to effects."[151]

Herbert Spencer expresses this view most succinctly in his 1862 First principles, "We come down then finally to Force, as the ultimate of ultimates . . . Space, Time, Matter, and Motion are apparently all necessary data of intelligence, yet a psychological analysis shows us that these are either built up of, or abstracted from, experiences of Force. Matter and Motion, as we know them, are differently conditioned manifestations of Force . . . a resistance we are to symbolize as the equivalent of the muscular force it opposes. In imagining a unit of matter we may not ignore this symbol, by which alone a unit of matter can be figured in thought as an existence. It is not allowable to speak as though there remained a conception of an existence when that conception has been eviscerated—deprived of the element of thought by which it is distinguished from empty space. Divest the conceived unit of matter of the objective correlate to our subjective sense of effort, and the entire fabric of physical conceptions disappears."[152]

Whatever the root of ontology physics seeks to describe turns out to be, it must be so staggeringly beautiful in its function as to connect everything we experience and know, from electrons to galaxies, and all beings, conscious feelings, and information in between. Its concept must represent a fundamental principle which can be described in words and math, but which can only be truly known in the experience of being it—yielding a human being the indescribable quality of emotional experience had only in living as but one representative of its consciousness. It must be so unique an idea as to give insights into the nature of a human's experience of love, sex, birth, death and every other forever-differently-describable sensation of consciousness. Its diversity must be capable of fulfilling the infinite grandeur and eternal wonder of that which is forever, and everything.

What could possibly be such a wondrously vast and creative agency as this?

One way to attempt to identify this kind of most common denominator—irreducible by further mechanical description—in our study of Nature is to consider how the most ruggedly demonstrated concepts of physics can be expressed in terms of each other. In an effort of integration in the trajectory of unity, a candidate tree of physical definitions can be prepared through the use of the entirely logical process of term substitution. In logic and computer science, complex expressions of ideas can be mapped in trees in possible configurations of definition, with each branch establishing its function as used in the aggregate definition. With root apparently outside of our present reach, a superior tree of natural philosophy is one that is capable of describing basic concepts of existence, framing extensive branches of knowledge across disciplines of observation. For the root of such a tree we would presumably seek the simplest fundamentals within an absolute frame of explanation for phenomena. However outrageously audacious such an architecture may sound, this is indeed what physics seeks, and where many of the greatest of scientists have believed hypothesis and test will ultimately lead. For those who understand the significance of this eternal physical mystery, this can be described fairly as a sacred quest for truth, wherever and however such truth may rightly be found.

So with a concept of coherence in mind driven by the sweep of observation, one can follow a path that brings definitions of physics together into expressions with fewer and more clearly defined kinds of parts. Success in such an exercise can be measured in the degree of exposition of 'leaf' behavior afforded in terms of 'root' concept. For the sake of open-minded discussion, let us examine how this process might be applied to our attempt to comprehend the nature of force. Consider how term substitution can be employed to articulate our comprehension of the basic physics concepts of electric and magnetic forces. Adapting definitions from the Oxford Dictionary of Physics (you can use any physics textbook you wish to reconstruct this exercise on your own), let's examine shorthand labels that define physics' mental picture of these basic *forces of Nature:*

## *Electric force:*

A [force]
    agency responsible for a body's [acceleration]
        change in [velocity]
           the [vector]
                quantity with linear direction
           and magnitude
           of [motion]
                changing of position relative to
    [time]
                  the enumeration of
        spatial cycles of position
      that [accelerates] [charges]
           [electric force] of a body giving rise to repulsion
           among bodies with like [electric force], and attraction
           among bodies with unlike [electric force]

Looks ugly, doesn't it? But pause here, for a moment. Re-read and think about these words carefully. These are the words that represent our very meaning of the phenomena in question. In other words, they are concepts that describe the primary mental picture we have of what is actually happening. This method represents a very interesting way for the philosophically inclined to enhance their appreciation for what is going on among the bodies forming all matter in the Universe that we know of. Notice the circularity in the definition of electric force. This is an important clue to one of the central points of this exercise.

Employing this tree, we can propose another way of expressing the definition of electric force, doing no harm to the scientific concept: electric force: an agency responsible for repelling bodies—made of the same agency—with similar property in agency and attracting bodies with opposite property in agency.

## *Magnetic force:*

A [force]
> agency responsible for a body's [acceleration]
of attraction or repulsion extant between [charges]
> [electric force] of a body giving rise to repulsion among bodies with like [electric force], and attraction among bodies with unlike [electric force]
in relative [motion]
> changing of position relative to [time]
> > the enumeration of spatial cycles of position

Here again, let us explore refinements in description, doing no harm to meaning. *Magnetic force: an agency of attraction and repulsion resulting from the relative motions of bodies of the agency labeled electric force.*

The somewhat dense discussion of philosophical integrations demonstrated above, and other integrations I didn't make the effort to articulate, may leave the adept student with some interesting insights, and ideas for further research
.

First, the simple and transcendental roles of certain concepts in our descriptions of Nature leap out at us. It becomes somewhat easier to identify and distinguish entities, movements, interactions, and dependencies. But the process also reveals true limits of our mental picture of Nature. It may surprise the philosopher that at least four seminal words employed within the definitions above are not included in the Oxford Dictionary of Physics: body, agency, attraction, and repulsion. These are, in fact, key concepts whose lack of further definition or declaration as fundamental contributes to the problems of physics today. Consider the word body. Of course, we all have a fairly straightforward idea of what is meant by body: an entity coexistent and distinct. But as we think about it, what do we really know an actual particle of 'charge' to be, in terms of physics, other than (1) exhibiting and responding to electromagnetic force, and (2) exhibiting inertial resistance . . . both within certain geometries of motion?

Consider then the intrinsic qualities of the ubiquitous word agency as the definition of the essence of force. Note that it is a term used recursively in the definition of electric force—meaning that the term is being used to define itself. What this is telling us is likely that force—the omnipresent agency forming everything we know, characterizes the fundamental. The amazing thing to consider is that it is quite possibly the ontological limit in principle of mechanical physics. By the fact of its basis as the relation governing all of our measures of physical interactions, force is a fundamental concept seemingly impenetrable to empirical inspection except through its own employment. In other words, it takes force to inspect force; it takes force to know force—at least within the framework of physics we currently can express.

Such an 'agency' appears to this student as perhaps the most basic concept in physical science, and therefore does not appear to be reducible beneath itself using available concepts of physics. Leaping to explanations of the forces of nuclear electrodynamics in terms of 'particle exchange' (facilitating intra-nuclear accelerations) and 'warped dimensions' (facilitating extra-nuclear accelerations) might simply represent the injection of levels of indirection and complexity in our study of the nature of matter in motion in medium.

Equipped with alternative concepts of space and time speculated above, we may follow the lead of prominent natural philosophers and conjecture that, indeed, 'force' represents a now-mathematized symbol of an ontological fundamental of physical reality. We gain confidence in this conjecture through two straightforward steps of further reasoning.

Firstly, in connection with a more refined concept of time, we can see that 'force' as symbolized in equations of action is ontologically incomplete; it lacks appreciation for the nature of time as cyclical motion. If the dimension of time is viewed as simply a dimension of measurement of ever-continuous motion, then, perhaps ironically, the positivist's appraisal of 'force' is all too accurate: the symbol is merely an instantaneous (infinitely partial) mathematical relation. The symbol does not capture the ontological significance of the process it attempts to map—the complete, unending ontological dynamism comprising bodies and their relations. The dimensions of 'force' do not map the territory of reality, because the qualities attributed to 'force' and 'body' are never intrinsically instantaneous—they are never empirically observable apart from paths of interaction. A simple analogy might make this point clearer: physics' concepts of 'force' (and 'mass') are to their underlying reality like 'photograph' is to 'movie'. The concept of time as motion makes it clearer how physics' ontological concepts of 'force' (and 'mass') presently express only mathematical snapshots of the reality they attempt to depict.

The underlying territory which the symbol of 'force' is intended to describe can only be mapped accurately on a canvas framed in terms the dimensions of the territory itself. Thus, the symbol of 'time' (perhaps as redefined above) must be combined with the symbol of 'force' if we wish to comprehend the true nature sketched by either symbol. The vector quantity of force which accomplishes this—however imprecise—is impulse: force x time. By this reasoning it is fair to assert that no 'force' in nature is really 'force'; ontologically speaking, we ought think of all instances of force as 'impulse' . . . or more simply, the active pressure gradient that appears as 'attraction' or 'repulsion'. Thus, gravity is not a 'force', it is an active pressure gradient. Nuclear 'forces' are not 'forces', they are active pressure gradients. The existential character of 'active pressure' is observable in its expression in bodies and their surrounding medium, yielding patterns of interaction within and among bodies, all describing equilibrium geometries in and of itself.

Secondly, from this perspective, we can see that 'force' and even 'impulse' as symbolized in equations of action may not express the total operational territory which they attempt to map. It should be obvious that a vector quantity of 'force' or 'impulse' can represent only a linear descriptor of a relationship or interaction, respectively. If the volumetric dimensionality we call 'space' is in fact a medium of ongoing pressure of an omnidirectional or more complex dynamic geometry, then linear measures of interactions within it surely yield only partial, approximate descriptions of the actual interactions.

For example, how sure are we that the application of impulse to matter creates acceleration where there was none before the application? Might it be, as implied in our discussion above concerning the nature of time, that the application of external impulse to matter instead alters the pattern of continuous acceleration within the fundamental particles of the matter with respect to the medium in which they move? Might it be that external impulse simply alters the pattern of helicity of fundamental bodies in motion?

If so, might this also give us insight into the nature of the asymmetry in the medium which we call gravity?

## Mass

As I've described previously, physics' concept of mass is inextricably entwined with its concept of force, and thus if the concept of force is eliminated by principle of the space-time warp advocated by the currently-prevailing, extreme wing of positivist philosophy, then the elimination of the concept of mass cannot be far behind. Whitehead succinctly captures the crucial point: "We obtain our knowledge of forces by having some theory about masses, and our knowledge of masses by having some theory about forces." [153]

Thus, from our speculation on the ontological irreducibility of force above, at least within the descriptive limits of the epistemology of present-day physics, it follows that mass possesses a similar noneliminability. Tyndall interprets Faraday's view on this assertion thusly: "What do we know of the atom from its force? You imagine a nucleus which may be called a, and surround it by forces which may be called m; to my mind the a or nucleus vanishes and the substance consists of the powers m. And, indeed, what notion can we form of the nucleus independent of its powers? What thought remains on which to hang the imagination of an a independent of the acknowledged forces?" Schelling expresses the same essentially Kantian notion from earlier philosophies in saying, "It is mere delusion of the phantasy that something, we know not what, remains after we have denuded an object of all the predicates belonging to it."[154]

In reexamining the nature of mass, let us start by revisiting the words of Newton quoted previously: "A body, from the inert nature of matter, is not without difficulty put out of its state of rest or motion. Upon which account, this vis insita may, by a most significant name, be called inertia (vis inertiae) or force of inactivity. But a body only exerts this force when another force, impressed upon it, endeavors to change its condition; and the exercise of this force may be considered as both resistance (resistentia) and impulse (impetus); it is resistance so far as the body, for maintaining its present state, opposes the force impressed; it is impulse so far as the body, by not easily giving way to the impressed force of another, endeavors to change the state of that other."

I suggest that the true nature of mass has been hiding in plain sight within these words for over three centuries. Can you see where the concept may be found? Consider a definition of inertia paraphrased from the Oxford Dictionary of Physics: the resistance of a body or system of bodies to acceleration.

If you turn over these words in your mind for a while, two things may become clear: (1) mass is resistance, purely and simply, and (2) this resistance is undefined unless it is conceived in the context of time (perhaps as redefined above)—it is unobservable except over paths of motion. Resistance is defined in physics as a force, and force expressed in terms of time is known as impulse. Since impulse, as formally defined, carries with it a direction of action, and since whatever the resistance 'mass' represents has no relative direction unless acted upon relatively, we might approximate the nature of mass to a system of continuous impulse which reveals a reactive directional vector only when relatively accelerated. In my view, such is a truer ontological concept of 'mass' (and more generally and interestingly, momentum)—a kind of coherently-sustained, dynamic system of ongoing pressure. When such a dynamic coherence—a 'body'—is relatively accelerated through the medium we hypothesized above, whose ontological nature is conjectured to be the substance of the bodies it makes and conveys, a relative resistive impulse is observed, inertial mass.

In this notion of mass, the temporal experience we sense and otherwise refer to as 'the flow of time' finds its ontology in a kind of pressure, whose action we observe in the dynamic coherence and perpetual acceleration of all elementary bodies—the antecedent process of mass. This concept is thoroughly consistent with our common sense of everyday interactions with objects: we push and pull, turn and twist, and compress and expand them. In all cases, we feel some geometrizable pressure in response. From whence does this pressure arise if not from the intrinsic dynamic nature of the bodies and the medium in which they exist? And in examining the nature of 'nuclear forces', why need we characterize the fundamental relations involved within 'mass' as arising from 'particle exchanges' somehow 'carrying' 'force', when we can more economically view 'force carrying particles' as simply density geometries in a medium whose intrinsic nature is otherwise undifferentiated, continuing pressure?

What, for God's sake, is a 'particle' if not a coherent, sustained geometry of 'pressure'? Surely, a particle cannot seriously be defined to be merely a geometric—or even geometrodynamical—boundary condition absent some notion of 'the grip of force' to make the boundary physically meaningful. If this assertion is true, then why need we employ any concept of particle beyond the notion of sustained density patterns in a universal, underlying field of pressure?

We can complete this mental motion picture by integrating our proposed concept of time into this proposed concept of mass. Consider our hypothetical example of an electron following (ideally) a helical trajectory through the medium. Recall that its hypothesized velocity is the ratio between the linear measure of distance traversed and the length of the nonlinear path actually followed. A linear motion of mass averaged from a nonlinear mass dynamic must represent some system of pressure gradients in equilibrium relative to the path of motion in a medium. Thus, very much as Aristotle intuited, velocity of mass—'uniform straight-line' motion—may well involve the continuing action of 'force'. Since we defined acceleration as the change in the ratio between the linear distance traversed and the nonlinear length of the path actually followed, divided by the nonlinear length of the path actually followed, the resistive impulse observed macroscopically as inertial mass represents the effort involved in changing the microscopic nonlinear geometry of motion of moving charges. In other words, if we presume that the intrinsic motion of fundamental bodies is always nonlinear, whereby 'straight-line velocity' is never but an average of motion, then the resistive impulse of inertia does not arise from acceleration, rather it arises from a change in the geometry of ongoing acceleration—somewhat like the impulse of resistance yielded when one attempts to alter the orientation of a spinning gyroscope, but in this case equally reactive to relative impulse regardless of the direction of application.

In considering the implications of this idea for our understanding of the true nature of inertial motion, it is interesting to note that, assuming every impulse is opposed by an equal and opposite impulse (which assumption I shall examine in future writings), there is one geometry of motion in which the ongoing application of an impulse of constant magnitude yields both microscopic constant acceleration and the macroscopic appearance of constant velocity: the helix . . . whose two-dimensional projection happens to be a sinusoidal wave.

# Chapter 7
# Speculations on Implications to Physics

We will speculate that that which exists is an infinite, eternal medium of continuous, omnidirectional pressure, sustaining coherent density geometries (impulses and bodies) thereof in dynamic equilibrium. For lack of an existing nomenclature free of conceptual baggage, let us refer to the hypothesized medium as potentum, to field gradients and impulses as potentia and matter particles as potentons. We can subjectively measure the scales and cycles of potentia and potentons only in terms of each other—relative measures using rulers and clocks made of same—and thus the observable Universe represents the maximally objective frame of reference by which comparative dimensional measurements may be conceived.

If true, the implications of this ontology are far reaching. Let us explore a few of these implications on specific questions of physics, and more broadly, science.

## Quantum Mechanics

Among the unsolved mysteries of physics, three stand out, all of them surfaced in the theory of quantum mechanics. The first is the strange correlation of properties of particles across arbitrary distances, which properties appear to be simultaneously determined for systems of particles when just one particle is observed. This seems to violate our common sense of causality. The second mystery is the Heisenberg uncertainty principle, which asserts an irreducible "fuzziness" among measurements of properties of particles.

The third, related mystery is quantization in action—or step-wise, rather than smooth—interactions among particles and radiation.

Let's tackle these questions in reverse order. The work of physicist David Hestenes has revealed a heretofore unrealized geometry hidden in Dirac's equations of electron motion: the helix. His work directly addresses basic questions of the physical meaning of spin and other quantum parameters. Although he is appropriately conservative in his work, it is nonetheless suggestive of the idea that the quantization of action in electromagnetic and particle interactions results from the fact that the fundamental motion of electrons is not linear or even stochastically so, but rather is predictably nonlinear, varying (stochastically or not) upon the idealized helical form. He suggests, as others have elaborated more recently[155], that the electrodynamics of self-interaction of charges, or the interaction of charges with the energetic medium, may give rise to this fundamental form of motion. It is not hard to extrapolate how this insight may shed light upon the nature of quantization of action in physics. For if fundamental, pristine motion of charges is helical (or if charges are vortices which express such a pattern of moving force), then we may come to see that "turns of the helix" offer a physical resolution of the dilemma between corpuscular and continuous notions of the substance of the medium, and, thereby, resolve some aspects of the mystery of quantization of action. Hestenes has gone on to build upon and expand the 19th century work of Clifford to establish a new, unifying branch of geometric mathematics called geometric algebra, which serves as a remarkably powerful tool not only to express quantum mechanical phenomena within this hypothesized framework and give its equations physical meaning, but to provide a genuine lingua franca for mathematical (and therefore dimensional) geometry as a whole.

To the second issue—the nature of the Heisenberg uncertainty principle—I will haphazardly speculate that the reason we cannot measure both position and velocity with deterministic precision rests in the centuries-old mathematical legacy of assumptions among physicists that inertial motion follows a straight line, rather than a nonlinear form. Consider that physicists' measurement of position of a particle can never align with their measurement of its velocity if the particle is always circling about the line average presumed by physicists to define its true momentum vector.

In other words, consider that the Heisenberg uncertainty principle is built-in to the equations of physicists' assumptions: it may be related to the radius of the helix of fundamental particles in motion which, thus, never occupy positions along their presumed vectors of momentum. Thus, this "irreducible fuzziness" may have nothing to do with intrinsic, idealized uncertainty in the motions and momenta of particles, but rather may result from a here-to-fore irreducible approximation built into equations of physics, an approximation resulting from the natural consequences of knowledge acquired from an "outside in" process of experimental examination. It seems likely that ultimate answers will most likely emerge from "inside out" interpretations of experimental data.

And now to the first and most perplexing problem of quantum mechanics—apparent non-local interaction of correlated fundamental particles: I will again speculate that the central feature of Nature missing from the epistemological and ontological framework of quantum mechanics is the conception of space as a dynamic, energetic medium. Consider the possibility that the reason fundamental particle systems appear to be non-locally correlated is that the total "system" includes an active, causal medium comprising every aspect of the subjective observer, the object observed and the experimental environment that envelopes them.

That is, consider that observation-driven correlations appear non-locally triggered not because of instantaneous communication between particles, but rather because the entire environment preceding and surrounding any experiment—including all experimental apparatus and the experimenters attending thereto—establishes a system-wide causal context that perfectly determines the outcome later allegedly "freely chosen" by the experimenter. If, in fact, the configuration of the experimental environment establishes patterns of ongoing influence in the medium sustaining and/or comprising every particle therein, and if the causal Universe thus even "knows" (causes, or even tends toward causing) the experimenter's choice in advance, then it is not really surprising that the experimenter observes "instantaneous" correlations among elements within the system, particularly if the experiment maintains a low noise floor.

The quest by experimenters to reduce the noise floor in quantum correlation experiments thus might become seen as simply increasing the ability of the experimenter to see selected, deterministic, causal relations already established across space and time.

Consider an analogy. If one were to construct a system to measure the correlation of the rolling motion of boats in an ocean near the two jetties of an inlet to a bay, it seems quite clear that—given the dynamics of an ocean—the establishment of the jetties, the sending and positioning of the boats and any detection apparatus to measure the motion thereof will establish a priori a total ocean wave superposition environment that cannot be separated from the results of "correlation" experiments on the rolling motion of the boats. In other words, if it is true that the medium of space is dynamic and energetic, it may be impossible in principle or in practice to construct an experimental framework that permits the experimentalist to separate him/herself or the testing apparatus from the establishment—indeed with noise reduction, the near perfection—of the means to reveal deterministic relations among particles in/of the medium.

This rather startling hypothesis suggests an ultimate irony in the conclusions drawn in the 20th century by quantum theorists: the exactness of instantaneous correlation among fundamental particles may represent a first-class case for large-scale—perhaps all-encompassing—determinism in Nature, directly at odds with the assumed ontological implications of Heisenberg's uncertainty principle.

The implications of this assertion to concepts of free will of conscious beings are beyond the scope of this volume, but there are some surprisingly interesting mechanisms by which a form of free will in conscious beings can emerge within this structure, and they will be explored in future writings.

COMING SOON TO A PAGE NEAR YOU:

BOOK TWO, DECEMBER 2009:

**EVOLVING ECONOMICS**


BOOK THREE, SUMMER 2010:

**2050**

**A DAY IN THE LIFE**


**A PREVIEW**

# The Science of Ontology and the Society of Science

"My message in this book, which now
draws to a close, is that if you believe the
experts when they tell you your naïve wit and critical
sense are worthless then in your own case you prove
them right . . . but if you resist their browbeating
there is hope in your case for individual salvation . . .
Such personal salvation can lie only in your in-born
assets of common sense (so dreadfully uncommon) and
good judgment. May those assets continue to
resist all scientific definition, codification,
quantification, and computerization . . . and though they
have now (in a world flattened under the heel of idiocy)
only a negative Darwinian survival value—may they
never be altogether bred-out of our remarkable species."[156]

Thomas Phipps

As the reader diligent enough to traverse carefully the words of this monograph may now intuit, the stakes for obtaining the right answers to the questions herein posed are staggeringly high.

The next book in this trilogy will focus on philosophy, science and the society possible if new insights revealed in physics are properly tapped. They include:

- string theory is a mistaken approach to a good idea: namely, all fundamental particles share a basic character; it is not that space-time is warped through 10-11 dimensions; the commonality is better realized as the intersection of helical motion in many directions;
- acknowledgement that the language of physics—indeed language in general—must evolve to reflect increasing comprehension of a

unified ontological root, revealing that words are symbols for diverse expressions of a unitary physical reality;

- realization that "information" and "physical pattern" are likely synonymous;
- discovery that remarkable non-local effects of consciousness may soon be increasingly comprehensible by physical science;
- recognition of an imminent dawn of staggering scientific and technological discovery; we are at the edge of an infinite frontier;
- potential implications of a unified ontology of physics to world spiritual traditions.

~

As a forward to forthcoming volumes, I will quote the greatest physicist ever to live, so far as I know. Sir Isaac Newton long ago asserted:

"The main business of natural philosophy is to argue from phenomena without feigning hypotheses, and to deduce causes from effects, till we come to the very first cause, which certainly is not mechanical; and not only to unfold the mechanism of the world, but chiefly to resolve these and such like questions. What is there in places almost empty of matter, and whence is it that the sun and planets gravitate towards one another, without dense matter between them? Whence is it that nature doth nothing in vain; and whence arises all that order and beauty which we see in the world? To what end are comets, and whence is it that planets move all one and the same way in orbs concentric, while comets move all manner of ways in orbs very eccentric, and what hinders the fixed stars from falling upon one another? How do the motions of the body follow from the will, and whence is the instinct in animals? . . . And these things being rightly dispatched, does it not appear from phenomena that there is a being incorporeal, living, intelligent, omnipresent, who, in infinite space, as it were in his sensory, sees the things themselves intimately, and thoroughly perceives them; and comprehends them wholly by their immediate presence to himself?"[157]

# Endnotes

1   Huygens, C. New Conjectures Concerning the Planetary Worlds, Their Inhabitants and Productions, c. 1690

2   Greene, B. The Elegant Universe, p.3-4, Vintage Books, 1999

3   Jammer, M. Concepts of Mass in Classical and Modern Physics, p. 224, Dover Publications, 1997

4   Jammer, M. Concepts of Space: The History of Theories of Space in Physics, p.9, Dover Publications, 1993

5   Jammer, M. Concepts of Space: The History of Theories of Space in Physics, p. 11, Dover Publications, 1993

6   Aristotle, as quoted in Jammer, M. Concepts of Space: The History of Theories of Space in Physics, p. 19, Dover Publications, 1993

7   Jammer, M. Concepts of Space: The History of Theories of Space in Physics, p.80, Dover Publications, 1993

8   Newton, I. Principia, p. 6

9   As quoted in Jammer, M. Concepts of Space: The History of Theories of Space in Physics, p. 109, Dover Publications, 1993

10  Duncan, G.M. The philosophical works of Leibnitz, p. 60, New Haven, 1890

11  A Collection of Papers which passed between Mr. Leibnitz and Dr. Clarke, p. 205, London, 1717

12  Jammer, M. Concepts of Space: The History of Theories of Space in Physics, p.127, Dover Publications, 1953

[13] There remains far less controversy than there ought be concerning the question of the relative merits of Ampere versus Maxwell formulations of electrodynamics. The former asserts a hitherto unacknowledged force relation between infinitesimal current elements that, if empirically validated, may hold significance for the theoretical apparatus of modern physics, and may suggest technological innovations of importance.

[14] Miller, D. The Ether-Drift Experiment and the Determination of the Absolute Motion of the Earth, Reviews of Modern Physics, Volume 5, July 1933, p. 203242

[15] Society for Scientific Exploration, "'THE EXPLORER" Summer 2004 (V20 N3 p17)

[16] Jammer, M. Concepts of Space: The History of Theories of Space in Physics, 139-140, Dover Publications, 1953

[17] As quoted in Davies, P. About Time, p. 23, Orion Productions, 1995

[18] As quoted in Davies, P. About Time, p. 72, Orion Productions, 1995

[19] As quoted in Davies, P. About Time, p. 72, Orion Productions, 1995

[20] As quoted in Davies, P. About Time, p. 267, Orion Productions, 1995

[21] As quoted in Davies, P. About Time, p. 104, Orion Productions, 1995

[22] Reichenbach, H. The Philosophy of Space and Time, p. 109, Dover Publications, 1957

[23] The Works of Plato by B. Jowett, vol. 3, p. 456, Oxford University Press, 1892 [24] Davies, P. About Time, p. 24, Orion Productions, 1995

[25] As quoted in Davies, P. About Time, p. 24 Orion Productions, 1995

[26] As quoted in Davies, P. About Time, p. 25 Orion Productions, 1995

[27] As quoted in Davies, P. About Time, p. 25 Orion Productions, 1995

[28] As quoted in Davies, P. About Time, p. 25-26 Orion Productions, 1995

[29] As quoted in Davies, P. About Time, p. 28 Orion Productions, 1995

[30] Greene, B. The Elegant Universe, p. 37, Vintage Books, 1999

[31] Jammer, M. Concepts of Mass in Classical and Modern Physics, p. 1, Dover Publications, 1997

[32] As quoted in Jammer, M. Concepts of Mass in Classical and Modern Physics, p. 2, Dover Publications, 1997

[33] As quoted in Jammer, M. Concepts of Mass in Classical and Modern Physics, p. 15, Dover Publications, 1997

34  Jammer, M. Concepts of Mass in Classical and Modern Physics, p. 17, Dover Publications, 1997

35  Jammer, M. Concepts of Mass in Classical and Modern Physics, p. 20, Dover Publications, 1997

36  Jammer, M. Concepts of Mass in Classical and Modern Physics, p. 26, Dover Publications, 1997

37  Jammer, M. Concepts of Mass in Classical and Modern Physics, p. 32, Dover Publications, 1997

38  Jammer, M. Concepts of Mass in Classical and Modern Physics, p. 32-33, Dover Publications, 1997

39  Jammer, M. Concepts of Mass in Classical and Modern Physics, p. 35-36, Dover Publications, 1997

40  Jammer, M. Concepts of Mass in Classical and Modern Physics, p. 39-40, Dover Publications, 1997

41  Jammer, M. Concepts of Mass in Classical and Modern Physics, p. 45-48, Dover Publications, 1997

42  As quoted in Jammer, M. Concepts of Mass in Classical and Modern Physics, p. 55, 57, Dover Publications, 1997

43  Sir Isaac Newton's Mathematical principles of natural philosophy and the System of the world, Motte's translation revised by F. Cajori, p. 1, University of California Press, Berkeley, 1947

44  Jammer, M. Concepts of Mass in Classical and Modern Physics, p. 65-67, Dover Publications, 1997

45  Jammer, M. Concepts of Mass in Classical and Modern Physics, p. 77-78, Dover Publications, 1997

46  Jammer, M. Concepts of Mass in Classical and Modern Physics, p. 84, Dover Publications, 1997

47  Jammer, M. Concepts of Mass in Classical and Modern Physics, p. 91-100, Dover Publications, 1997

48  Jammer, M. Concepts of Mass in Classical and Modern Physics, p. 141, Dover Publications, 1997

49  Jammer, M. Concepts of Mass in Classical and Modern Physics, p. 142, Dover Publications, 1997

[50] Jammer, M. Concepts of Mass in Classical and Modern Physics, p. 151-152, Dover Publications, 1997

[51] Jammer, M. Concepts of Force, p. 17, Dover Publications, 1999

[52] Jammer, M. Concepts of Force, p. 19, 23, Dover Publications, 1999

[53] Jammer, M. Concepts of Force, p. 27-30, Dover Publications, 1999

[54] Jammer, M. Concepts of Force, p. 34-35, Dover Publications, 1999

[55] Jammer, M. Concepts of Force, p. 28, Dover Publications, 1999

[56] Jammer, M. Concepts of Force, p. 51, Dover Publications, 1999

[57] Jammer, M. Concepts of Force, p. 62, Dover Publications, 1999

[58] Jammer, M. Concepts of Force, p. 53, 70, Dover Publications, 1999

[59] Jammer, M. Concepts of Force, p. 86, Dover Publications, 1999

[60] Jammer, M. Concepts of Force, p. 86, Dover Publications, 1999

[61] Jammer, M. Concepts of Force, p. 94, Dover Publications, 1999

[62] Jammer, M. Concepts of Force, p. 100-102, Dover Publications, 1999

[63] Jammer, M. Concepts of Force, p. 104-105, 108, Dover Publications, 1999

[64] Jammer, M. Concepts of Force, p. 116-119, Dover Publications, 1999

[65] Jammer, M. Concepts of Force, p. 119-120, Dover Publications, 1999

[66] Jammer, M. Concepts of Force, p. 121,125, Dover Publications, 1999

[67] As quoted in Jammer, M. Concepts of Force, p. 134-135, Dover Publications, 1999

[68] Jammer, M. Concepts of Force, p. 139, Dover Publications, 1999

[69] Jammer, M. Concepts of Force, p. 140, Dover Publications, 1999

[70] Jammer, M. Concepts of Force, p. 160-162, Dover Publications, 1999

[71] Jammer, M. Concepts of Force, p. 180-181, Dover Publications, 1999

[72] Jammer, M. Concepts of Force, p. 182, Dover Publications, 1999

[73] Jammer, M. Concepts of Force, p. 204-221, Dover Publications, 1999

[74] Jammer, M. Concepts of Force, p. 145-146, Dover Publications, 1999

[75] Jammer, M. Concepts of Force, p. 225, Dover Publications, 1999

[76] Jammer, M. Concepts of Force, p. 204-221, Dover Publications, 1999

[77] Jammer, M. Concepts of Force, p. 187, Dover Publications, 1999

[78] Jammer, M. Concepts of Force, p. 221, Dover Publications, 1999

[79] Forward to 1953 edition of Jammer, M. Concepts of Space: The History of Theories of Space in Physics, p. xvii, Dover Publications, 1953

80  Phipps, T. Heretical Verities, Urbana, 1986

81  As quoted in Lightman, A. Great Ideas in Physics, p. 135-136, McGraw-Hill, 2000

82  Lightman, A. Great Ideas in Physics, p. 137, McGraw-Hill, 2000

83  As quoted in Davies, P. About Time, p. 53 Orion Productions, 1995

84  Greene, B. The Elegant Universe, p. 66, Vintage Books, 1999

85  Greene, B. The Elegant Universe, p. 57-58, Vintage Books, 1999

86  It is important to note that gravitational force and inertial resistance are, in fact, empirically distinguishable. The tidal forces resulting from the small angular curvature of a gravitational field allow the distinction to be measured. This fact may also hold significance for discovering an underlying electrodynamic action of gravitational force and inertial resistance.

87  John Wheeler, as quoted in Greene, B. The Elegant Universe, p. 72, Vintage Books, 1999

88  Greene, B. The Elegant Universe, p. 78, Vintage Books, 1999

89  Review McCausland, I. Anomalies in the History of Relativity, http://www.scientificexploration.org/jse/articles/mccausland/1.html, and Phipps, T. Heretical Verities, Urbana, 1986, and Silvertooth, E.W., Experimental detection of the ether, Speculations in Science and Technology, Vol. 10, No. 1, p. 3

90  As quoted in Davies, P. About Time, p. 51 Orion Productions, 1995

91  Zeilik, M. & Gaustad, J. Astronomy: the Cosmic Perspective, p. 147, Harper & Row, 1983

92  Lerner, E. The Big Bang Never Happened, p. 133, Random House, 1991

93  Lerner, E. The Big Bang Never Happened, p. 140-141, Random House, 1991

94  Ferris, T. The Whole Shebang, p. 205-206, Simon and Schuster, 1997

95  Lerner, E. The Big Bang Never Happened, p. 338-339, Random House, 1991

96  Lerner, E. The Big Bang Never Happened, p. 340, Random House, 1991

97  Phipps, T. Heretical Verities, p. 630, Urbana, 1986

98  Hawking, S. The Universe in a Nutshell, p. 21, Bantam, 2001

99  http://www.cnn.com/2001/TECH/space/04/02/hubble.images/

[100] Davies, P. About Time, p. 162, Orion Productions, 1995

[101] Davies, P. About Time, p. 106-107, Orion Productions, 1995

[102] Davies, P. About Time, p. 111, Orion Productions, 1995

[103] Davies, P. About Time, p. 124, Orion Productions, 1995

[104] Hawking, S. The Universe in a Nutshell, p. 186-197, Bantam, 2001

[105] http://www.msnbc.com/news/692184.asp

[106] http://www.space.com/scienceastronomy/astronomy/dark_galaxies_010105.html

[107] Lerner, E. The Big Bang Never Happened, Random House, 1991

[108] As quoted in Lerner, E. The Big Bang Never Happened, p. 125, Random House, 1991

[109] As quoted in Ratcliffe, H. The First Crisis on Cosmology Conference, Progress in Physics, December 2005

[110] Kaku, M. Visions, p. 343, Doubleday, 1997

[111] Kaku, M. Visions, p. 342, Doubleday, 1997

[112] Hawking, S. The Universe in a Nutshell, p. 142, Bantam, 2001

[113] Kaku, M. Visions, p. 352, Doubleday, 1997

[114] Greene, B. The Elegant Universe, p. 129, Vintage Books, 1999

[115] Greene, B. The Elegant Universe, p. 137, Vintage Books, 1999

[116] Greene, B. The Elegant Universe, p. 138-139, Vintage Books, 1999

[117] Greene, B. The Elegant Universe, p. 146, Vintage Books, 1999

[118] Greene, B. The Elegant Universe, p. 201, 203, Vintage Books, 1999

[119] Greene, B. The Elegant Universe, p. 222, Vintage Books, 1999

[120] Greene, B. The Elegant Universe, p. 266-269, Vintage Books, 1999

[121] Greene, B. The Elegant Universe, p. 280, Vintage Books, 1999

[122] As quoted in Greene, B. The Elegant Universe, p. 280, Vintage Books, 1999

[123] As quoted in Greene, B. The Elegant Universe, p. 280, Vintage Books, 1999

[124] Phipps, T. Heretical Verities, p. 630, Urbana, 1986

[125] As quoted in Phipps, T. Heretical Verities, p. 9, Urbana, 1986

[126] Arp, H. Seeing Red, Aperion, 1998

[127] Arp, H. Seeing Red, p. ii, Aperion, 1998

[128] See http://www.cnn.com/2002/TECH/space/01/08/hubble.universe/index.html

[129] Seehttp://astronomy.com/Content/Dynamic/Articles/000/000/000/445wrzvl. asp

[130] See Science News, Sept. 8, 2001 (vol. 160, #10)

[131] See Dr. Manuel's research at *http://www.umr.edu/~oml*

[132] Review Phipps, T. Heretical Verities, Urbana, 1986

[133] Review McCausland, I. Anomalies in the History of Relativity, http://www. scientificexploration.org/jse/articles/mccausland/1.html

[134] Phipps, T. Heretical Verities, p. 102-108, Urbana, 1986

[135] Dicke, R. Mach's Principle and Equivalence, p. 3, Proceedings of the International School of Physics Enrico Fermi, June 19—July 1, 1961

[136] Dicke, R. Mach's Principle and Equivalence, p. 1, Proceedings of the International School of Physics Enrico Fermi, June 19—July 1, 1961

[137] Dicke, R. Mach's Principle and Equivalence, p. 6-14, Proceedings of the International School of Physics Enrico Fermi, June 19—July 1, 1961

[138] Sakharov, A. Vacuum Quantum Fluctuations in Curved Space and the Theory of Gravitation, Soviet Physics-Doklady, 12(11), 1040-1

[139] Puthoff, H. Gravity as a zero-point fluctuation force, Physical Review A, Volume 39, Number 5, p. 2333

[140] Puthoff, H. Gravity as a zero-point fluctuation force, Physical Review A, Volume 39, Number 5, p. 2333

[141] Haisch, B., Rueda, A., Puthoff, H. Inertia as a zero-point Lorentz force, Physics Review A, 49, 678, 1994

[142] Rueda and Haisch, Physics Letters A, 240, 115, 1998; Foundations of Physics, 28, 1057, 1998143 See *http://calphysics.org*

[144] Haisch, B., Rueda, A. Geometrodynamics, Inertia and the Quantum Vacuum, AIAA/ASME/SAE/ASEE Joint Propulsion Conference, Salt Lake City, July 8-11, 2001145 See *http://calphysics.org/gravitation.html*

[146] Puthoff, H. Polarizable-vacuum approach to general relativity, p. 1, Gravitation and Cosmology: From the Hubble Radius to the Planck Scale, Eds. R. Amoroso, G. Hunter, M. Kafatos and J.-P. Vigier, Kluwer Academic Press, Dordrecht, the Netherlands, in press, 2001147 See http://calphysics. org/gravitation.html

[148] http://motionsciences.org/frameset.html?whereAmI=Research&menucol or=%23 E1C8C8&location=/research/index.html&whereAmIFlash=3

[149] Kuhn, T. The Structure of Scientific Revolutions, p. 6, University of Chicago Press, 1962

[150] Jammer, M. Concepts of Force, p. 264, Dover Publications, 1999

[151] As quoted in Jammer, M. Concepts of Force, p. iv-v, Dover Publications, 1999

[152] As quoted in Jammer, M. Concepts of Force, p. 186, Dover Publications, 1999

[153] Jammer, M. Concepts of Mass in Classical and Modern Physics, p. 120, Dover Publications, 1997

[154] As quoted in Jammer, M. Concepts of Force, p. 184-185, Dover Publications, 1999 [155] See the work of Ibison, M. available from Institute for Advanced Studies at Austin, and the work of Cole, D., Boston University

[156] Phipps, T. Heretical Verities, p. 630, Urbana, 1986

[157] Jammer, M. Concepts of Force, p. 152-153, Dover Publications, 1999

Made in the USA
Lexington, KY
27 January 2012